FOR THE WANT OF A SHOE, THE HORSE WAS LOST:

MAKING A MAN

CHRIS GREGORY

COPYRIGHT © 2023

For the Want of a Shoe, the Horse Was Lost

Making a Man

Print ISBN 978-0-9833140-5-9

E-book ISBN 978-0-9833140-6-6

The names of many have been changed in this work.

Book cover photo depicting mounted color guard, circa 1985, provided by the New Mexico Military Institute Archives, 2022.

Author Chris Gregory is on the left, Dick Butler holding the American flag, and Mike Gregory holding the New Mexico flag.

❀ Created with Vellum

DEDICATION

This book is dedicated with love to my folks, Mike and Diana Gregory, and my little brother, Mike Gregory. You all gave me such a childhood.

THE HORSE IS MADE READY FOR THE DAY OF BATTLE, BUT VICTORY RESTS WITH THE LORD.

PROVERBS 21:31 NIV

INTRODUCTION

You are holding the second installment in this series of memoirs. This one takes me from the very beginning in Fort Irwin, California to a Las Vegas, New Mexico, ranching childhood, then through three years as a cadet at New Mexico Military Institute in Roswell, New Mexico. These are the stories that produced the adult that I eventually became, for better or worse. Laugh, enjoy, and learn to avoid the mistakes I've made.

The following proverb, *For the Want of a Nail, the Shoe Was Lost*, penned by the great Benjamin Franklin in *Poor Richard's Almanac*, serves as the catalyst for these stories. This proverb, or a variation of it, has been used for many centuries. The gist has always remained the same; e.g., a horseshoe nail costing a kingdom.

For the want of a nail, the shoe was lost,
For the want of a shoe, the horse was lost,
For the want of a horse, the rider was lost,
For the want of a rider, the battle was lost,
For the want of a battle, the kingdom was lost,
And all for the want of a horseshoe nail.

Here I am simply copying the idea of one of my favorite authors, James Herriot, who wrote a series of books, titling each one from a line of the hymn written by Cecil Francis Alexander:

All things bright and beautiful,
    All creatures great and small,
    All things wise and wonderful:
    The Lord God made them all.

Those James Herriot books were some of my favorites, and my mimicry is only in the highest form of flattery and admiration.

TABLE OF CONTENTS

PART I

RANCH BOYS

1

FAMILY MATTERS

Oh, that my brother, Little Mike, would pick up a pen. My brother is one of the funniest people I have ever known, and if he were to transfer his stories to paper, his name would become a household one. When eating with him, I prefer to eat mashed potatoes or refried beans, because they are harder to choke on when you're laughing and trying to swallow at the same time.

Little Mike has a bright mind and is gifted with timing and execution when telling a story. I can see his gleaming eyes squinting as he tells a story, trying hard to get the words out while he is laughing with and at you. It is contagious and funny and great – as long as you are not the central theme of the story. The problem is that I am often the target, but I end up laughing too hard to defend myself. If you are the target, you must take it like a man. Your poorly timed defense gains no sympathy after the verdict has been reached by those choking and laughing around the table as Little Mike announces the sentence.

My brother and I were raised in and around Las Vegas, New Mexico. This was the New Mexico of yesteryear when cowboys still ruled the world and liberal politicians had not ruined a designated part of God's country. We had a great upbringing, although I didn't see it that way at the time. Comparing the relative ease of my sleeping-in, TV-watching, bowling, movie-going, town-dwelling friends, I thought ranch life was a hard sentence. How little a child knows about the important foundation being built under them while it is happening. Having such strong and farsighted parents meant that I was to learn work ethic despite my resistance. A person can fake a lot of things in life, but work ethic is not one of those things.

I watched an interview with a successful writer who commented, "A horrible childhood is the greatest wealth a writer can have." If that is the case, I am a bankrupt writer, for I was certainly fortunate to be born to the parents God gifted me with. Their story is a great one. They were in separate registration lines at New Mexico Highlands University and only glimpsed at each other without any words passing between them. Mom went home and told her mother that she had seen the man she was going to marry. This was not in line with the typical training and behavior of a young woman expected to become a Catholic nun, which was the plan for her at that time, so I am sure they were both surprised. My dad had noticed this little Hispanic girl and was taken with her, determined to meet her later. They were both right.

Our dad was a workaholic who came from a rough childhood. His life with an abusive mother and a hard-working, mining father started in the mountain mining towns of Colorado. By the age of fourteen, he was working alongside his dad in the uranium mines of New Mexico.

Our mom was a beautiful woman of Hispanic descent and was loved by all. Her ability to soothe my dad was quite possibly the only reason that I'm still around to put these words to pen. That I was a difficult child has never been contested. I had the bad habit of not learning from others' wisdom or mistakes. I would watch one of my friends get paddled at school for some offense then go do the same to see if I would suffer a similar fate. I was always surprised that those rules applied to me, as well. I still am, sometimes, although riding this big ball around the sun so many times has tempered my will.

Spoken warnings, whether delivered softly or at volume, did little to change my character. The stern application of a belt or similar device proved effective in the way that the Bible recommends. Had I been born to hippies, my last days would most likely have been spent behind bars and ended well before my twenty-fifth birthday.

My mother's parents, quite the characters in their own right, lived in town and were an important part of our lives. My dad's folks were not a big part of my youth.

My granny's name was Helen Lucero, but I called her Cholla. She worked in the same dentist's office for over sixty years and never learned to drive.

Cholla was the oldest of twelve or thirteen girls, as I remember, and she had been brought to Las Vegas, New Mexico, as a child in a horse-drawn wagon. From that landing, Las Vegas was her home forever more. She had a superstitious nature but was unable to instill that fear in me, much as she tried. Little Mike was not similarly immune. Stories of La Llorona seemed as made up to me then as now, and Cholla was exasperated by the fact that I would not take her seriously. Her obvious gusto and sincerity convinced Little Mike often, and he couldn't understand how I could laugh at such depressing and dangerous tales. She loved me through it anyway, as only a grandmother could.

My granddad was Cristobal Lucero. I called him Sambo, and he had quite a story himself. He came home from school one day in the fourth grade to find that his folks had sold him to a guy passing through who owned a sheep ranch. He was given new clothes for the first time in his life, a pair of overalls. He said he got rained on that first night he was out with the sheep, and the overalls shrank halfway to his knees. At least they were new. Being sold as a 10-year-old had to have had a psychological impact, but I remember Sambo as being a quiet, hard-working man with the uncanny ability to stay clean no matter what manner of work we had survived at the ranch. He would work as hard as any of us, but his clean white shirt would still be a clean white shirt at the end of the day.

For the record, I have always been one to fix a parent's mistake in naming their offspring, and often the nicknames I assign stick. Cholla and Sambo were living testimonies to my skill. While many have tried to hang a nickname on me, it has never seemed to have a lasting effect. Well done, Mom and Dad.

Sambo was a self-made man, and he owned a dry-cleaning business in the days when that was an important part of any progressive town. He and Cholla raised two natural daughters and another one they rescued from a lice-ridden crib she was lying in at one of Cholla's sister's houses. In those days, many Hispanic Catholic families had more children than they could afford, including Cholla's sister. Sambo and Cholla spoke more Spanish than English to each other, so Little Mike and I were not always privy to the gist of their conversations or arguments. Even when they fought and Cholla kicked Sambo out of the house, he would still drive back to her house to take her to work, get her groceries, pick her up from work, and drive her home every night. The sense of responsibility ran deep in that man.

We stayed with Sambo and Cholla on many occasions, and Cholla walked around the house armed with her *vaceta*. It was a leather thong, about three or four feet long with a loop tied in the middle to fit around her wrist, making the one thong into two whips. It could have been quite the little switch in the right hands, but Cholla wasn't built stout enough to impart much damage. If she caught you on bare skin, it could leave a nice welt that had more value at a later time to instill guilt in Cholla that I'm not certain she didn't fake anyway. I would often laugh while she chased me, which encouraged us both even more – her with anger, me with self-preservation and mirth. I know that I had to be such a tough kid to have around.

Although I first drew breath in California where my dad was an officer in the US Army Field Artillery, I don't remember a thing about Fort Irwin in Death Valley. From the stories, it seems an apt name for the place, and I am just as happy that I don't remember it. Dad went from the military in California to law school in Albuquerque, New Mexico, and my only memory of that city was a white dog that we had named Cotton. Other than that, Albuquerque remains as big a mystery of my past as Fort Irwin.

I do have some photos of my early years. Since I was the firstborn, the miracle of my appearance and my early conquests were documented. I think my little brother may have a picture of being born, one day at the park when he was ten, and then high school graduation. By the time the second child rolled around, the novelty had worn off. The fact that my behavior required all of my parents' attention had to play into that as well.

2

THE RED BELT

My mom was a model of beauty and obedience as a child, and I am certain that every one of her friends was compared to her by their own parents at that time. I can hear them now, saying, "Why don't you act more like Diana?" as youngsters were held to that magnificent standard.

Most of the Hispanic culture in Las Vegas was Catholic, my grandparents included. My mom was considering following the path to becoming a nun. We often teased her that she met my dad and became a "some" instead of a "none."

My mom has a younger sister, JoAnn. Not only was JoAnn compared to the incredible standard set by Diana, but she had to live seeing it every day. And for those who are a bit rebellious by nature, experience talking here, that can lead to behaving even worse.

I think my brother and I were comparable to them, with me being more like Aunt JoAnn and Little Mike being more like our mom. Anyway, Aunt JoAnn seemed to really get me and understand how my devious nature could only be directed and never controlled.

Aunt JoAnn left Las Vegas to become a stewardess in the years when that was a wonderful career that was only available to the prettiest and most glamorous young girls who applied. In the different

America of the time, being a stewardess automatically meant that you were, cute, fit, and knew how to flirt and giggle at the right time to get that businessman to buy another ticket on that airline. By all accounts, Aunt JoAnn was a gifted stewardess. Photos from those years tell the story of a gorgeous young Hispanic woman with a determined smirk evident in every look.

Aunt JoAnn doted on her nephews and sent gifts for every occasion. Since we were obviously cowboys of the first order, we each needed a set of pistols, which were an essential part of the uniform. I can still remember the pair of pistols and their white plastic grips, the finest imitation leather holsters, and a matching red belt made of a rubbery plastic that had been molded with a western tooling pattern.

"You boys need to clean your room," Mom said as she looked in on us.

"In a minute," I replied.

I was wearing my new pistol set and was clearly above the level of room cleaner. Besides, I was in the middle of a trade with Little Mike. This was still at a time when I could get him to give me a dime for a nickel since the nickel was bigger. Better yet, if I found a shiny copper penny, I could get a quarter out of him because the penny was obviously unique and more valuable. It only got that color from the trace amounts of gold in it, as every seven-year-old will explain to the six-year-old who has a smaller and silver-colored dime to trade.

"Right now," Mom said.

"We can't right now," I said. "Can't you see I'm busy?"

"I said right now," Mom repeated. An edge was coming into her voice that I had caused before and recognized easily. "We are leaving in twenty minutes and the room had better be clean."

"I told you we can't right now," I said.

Well, this kind of disrespect was not (and should not) to be tolerated. Gun-toting, money-trading kid or not.

"Take those guns off right now, young man," Mom said as she marched into our room. "You will do as you are told. Seven-year-olds do not talk to adults that way, ever."

I had done it. Little Mike was, as almost always, an innocent bystander who was about to learn another lesson about picking whom to hang out with. Not that he had that many choices in those days.

Since I could see the storm that I had caused coming, I reluctantly unbuckled my guns. Mom took the belt from my hands, and the holsters slid neatly off the end, leaving her with the perfect weapon to impart my badly needed lesson. Little Mike and I both got a spanking with that red belt, and it produced such exceptional results that it became reserved for that purpose. The tooling provided traction upon contact, and the construction material had that perfect weight and aerodynamics that made it better than any switch that could have been cut from the most supple of trees. Cholla had her *vaceta*, and Mom had the red belt.

I got to hear, "GO GET THE RED BELT," many times as a youngster. This is probably one of the main reasons that I get to live in my own house these days instead of one with bars and a roommate I didn't choose. I wouldn't have said so at the time, but that red belt was one of the best things that happened in my childhood.

3

WORK ETHIC LESSONS

After Dad completed law school, we moved back to Las Vegas, New Mexico. Dad started his law practice, and Little Mike and I started our adventures in growing up. Although we lived in a house on 7th Street in the early years, my main memories of being a kid have to do with the ranch.

The ranch was my dad's dream. It was 744 acres of mountain beauty located about twenty miles west of Las Vegas. Dad also leased the land behind it, and that sat up against the national forest, so we were the last place on the road before the wilderness.

That old house had some definite history. It was over one hundred years old in the 1970s. Hand-peeled logs supported the flat roof, and the adobe walls were over four feet thick. There was a front porch that led to the living room where the fireplace was. If you turned right, it led to two unheated bedrooms and a bathroom. My brother and I shared one bedroom and had separate beds. My folks shared the other one, which had only one bed. In the winter, a glass of water on the bedside table would have ice formed on top of it some mornings, and frozen pipes were the norm. If you turned left from the front door, you would be in the kitchen where there was a wood stove that Mom continued to use for cooking until after I had left home. The counter was multicolored tile. Avocado green, dandelion yellow, cotton candy pink, and similar colors fought a battle for superiority with no single color coming out the winner. That was it, but it was enough.

No matter which window you chose to look out of at night, the only lights that you would see were the ones that God had produced or the occasional beacon lights from airplanes. This fact saved the ranch one night when I saw an odd orange glow through the kitchen window and roused my folks and brother. The woodshed and the back of the Green Machine Jeep had caught fire. It was only ten feet or so from the back of the house, so our efforts and the arrival of the fire department prevented a catastrophe.

My folks bought the ranch from an old woman named Clare. She had killed a bear in the living room. He had come in the front door while she was cooking, so she went out the back door of the kitchen with a rifle, went around the house, then came back in the front door to end his harassment. I remember her sitting in the kitchen at a table that would have been more at home as a workbench in a mechanic's garage, telling the story as if the bear were a wasp and the rifle a flyswatter. It wasn't a big thing to her, which says a lot about that old woman.

The ranch had the ruins of several old adobe houses that had once been a little settlement known as San Pablo. This was right up Dad's alley because he wanted to build onto the main house. In hindsight, I'm glad those ruins were there. It would have been a lot harder to make the bricks by hand, and I am certain that it would have been on the agenda if those old houses hadn't been sitting there waiting for Dad to harvest the low-hanging fruit. Even in New Mexico, I never saw adobe bricks in the hardware store.

"Come on now, boys," Dad said as Little Mike and I were trying to release one more brick from its mud-chink prison where it had been entombed for maybe a century. "You've got to put some gusto into that crowbar."

Bear in mind that the bricks ranged from twenty pounds to forty pounds each. I was probably eight or so when this project was in full swing, and although I was tall for my age, I could have hidden behind the crowbar that we were trying to use. Little Mike was just that; he didn't grow much in height until he was close to twenty, so the two of us combined were not much good for the task at hand. The "gusto" Dad was referring to hadn't developed yet.

"We're only going to work another hour or so until breakfast," Dad said with a smile. "If we get this trailer loaded by then, we can unload it before lunch."

His enthusiasm was not infectious. Since Dad was a lawyer during the week, his vacation every weekend involved sweat and hard work on the ranch. When I asked him about an actual vacation one time, his answer was, "Son, you live in a place where a lot of people go for vacation. Just look at these beautiful mountains. Every day is like a vacation out here." Since he argued professionally, my protests came to naught.

Gregory men have always been early risers, and Dad would be up by 4:00 in the morning to get some office work done. Little Mike and I were allowed the luxury of our beds until 5:30 or so, and then we were out the door to get some work done around the place. It could be cutting wood, building fence, or as on this particular weekend, carefully dismantling an old adobe house to use the bricks on our house. We would work until 9:00 or so and then sit down to a wonderful breakfast. My mom could sure cook on that old wood stove, and those breakfasts were always amazing.

"You got it," Little Mike encouraged me as the brick came free. "I got this end, you get that one."

With that, we staggered under our load as we added another brick to the growing pile in the back of the flatbed trailer.

"Yep boys," Dad said, wiping the sweat from his brow. "We get another thousand or so and we'll have a real good start on an addition to the house."

Oh, that's all, I thought to myself. At least we had something to do. Little Mike and I picked up the crowbar and started on another adobe brick.

Although no kid appreciates it at the time, learning work ethic will serve them well throughout their life. It is not the sort of thing you learn by reading about it or being told about it. You must see it in those who raise you and you need to participate in it as well. Both of my parents were blessed with an abundant work ethic.

4

COUSINS

The ranch was a place of wonder for youngsters. One of Cholla's sisters had kids who were close in age to Little Mike and me, and they got to come to the ranch quite often. Rick was just a bit older than I was, and Gus was between Little Mike and me.

Our free time was spent building forts, climbing the mountain, exploring an old copper mine that was on the ranch, and finding all sorts of neat stuff left behind by the previous owners. At least, it seemed pretty neat to us kids.

"Here's a bullet," Rick said.

The New Mexico climate is kind to metal objects, and things don't rust or disintegrate as they do in more humid environments. The bullet was tarnished but otherwise still good.

"We should take it to Dad," Little Mike said.

"I think we should see if we can make it explode," Gus said.

Well, explode is a word that will get any preteen boy to smile and pay attention.

"Maybe we can smash it on a rock," I said.

With that, we all crouched around Rick's newfound treasure. It was centered on a flat rock, and we each took turns trying to smash it with a rock of our choosing. Trying to find a rock in that place was like trying to find water in the ocean. They were everywhere, in every shape and every size.

"That thing can sure take a hit," Rick said after it had been hit several times.

"The end of it is all bent up," I said.

"Maybe it has to be hit just right on the end," Little Mike said.

"Who knows," I said, swinging my rock one more time.

There was a muffled crack and the bullet disappeared, along with its casing. I now know that it was a rimfire .22 shell, but at the time I had no idea. My rock had finally set off the charge, and the bullet had exploded. Without a rifle to hold it, the shell probably went further than the slug, but both of them disappeared from sight.

"That was a letdown," Gus said.

"I expected it to be louder or have some flames and smoke," Little Mike said.

"I guess it was a dud," I said. "I wonder where it went?"

"Maybe it exploded into nothing between the rocks," Rick suggested.

"I guess so," I agreed.

None of us were shot, and the missing bullet and casing were never found, so once again, God had decided to let us continue to survive our childhood.

5

STAYING CENTERED ON A HORSE

My brother likes to tell the story of us riding horses on the ranch as kids. Since my version is being written, I hope it will become the official one. Would Winston Churchill call me the victor? Little Mike will embellish and enhance any version, truth or not, to the point that you will be crying at the telling while a little pee escapes into your drawers.

We grew up riding all the time. We mainly rode bareback, since saddles were heavy and awkward for a youngster, but always with a horse between our knees. Little Mike is sixteen months my junior, which is not a lot amongst adults but represents a great expanse when you are under the age of ten. He kept falling off his horse. This caused a couple of problems for me in that I would have to catch his now loose horse, stop and dismount my horse to help him on, then remount a bareback horse, which represented a pretty good challenge when your head barely comes up to the middle of the horse's ribcage.

Coming off a horse in those days resulted in very few of the injuries that come in later years when your body lands more like a brittle branch than a bag of ashes. There is also the factor of gross weight, and I was so skinny in those days that the impact was rendered harder by my aerodynamics than it was by gravity. We would even play games that involved jumping off the horses, which I'm certain later led to bulldogging becoming my chosen sport. But to have your brother come off time and again when it wasn't A: planned, B: part of a game, or C: handy; well, that was a problem. Luckily, I was a natural-born problem solver.

"Can't you stay on for just a few minutes?" I asked my brother, as he lay sprawled in the picturesque clover and grama of the pasture we had been riding through.

He looked up sheepishly as I led his horse back to him. "I'm sorry," was all he could manage.

"I don't see how you can come off with a saddle on," I said.

"He was bucking," he replied. "How was I supposed to stay on?"

"With a saddle," I said. "Just hold on next time."

"I'll try," he said.

I was disappointed by his efforts as I caught his horse again within minutes.

When the horses were saddled, falling off was inexcusable. As Little Mike continued the habit, I decided that there had to be a solution. You will sometimes hear people expand on the benefits and uses of duct tape. While they are not wrong, the truth is that baling wire can be an even more effective fastener than duct tape. I retrieved some pieces of baling wire, which were everywhere on the ranch, and wired Little Mike into the saddle. Wiring his belt to the saddle horn and his feet in the stirrups made Little Mike as sticky as any rodeo hero ever was. He continued to stay in the saddle and our riding going forward was much easier on me. Problem solved.

Little Mike brought up my brilliance to Dad some days later as he was bragging about staying on his horse better and how his big brother fixed his riding concern. Dad was less than impressed with my invention, and the practice of wiring Little Mike to a horse was immediately banned. No exceptions.

They say God watches over little children and idiots, and I was halfway between at the time. As luck would have it, no permanent physical problems had resulted from my inventiveness. However, there remains a mental scar inflicted on my brother from the experience. More than four decades have passed, but the wire holds.

6

DAD'S TRUCKS

Dad bought a short-bed Chevy pickup. It was dark green and cool looking, but it was two-wheel drive and not suited to getting back and forth to the ranch on at least ninety days a year. It used to rain in New Mexico in those days more than now, so the ten miles of dirt road that were part of the trip would become very hard to navigate. The New Mexico clay made for great bricks when mixed with straw but slippery as fresh cow patties mixed with water.

"How do you get the music out of these screens?" I asked my dad, referring to the speakers that looked like screens to me. Little Mike and I were sitting on the bench seat in the truck while Dad was driving to town from the ranch.

"There is a little band under the dashboard that plays it," Dad answered.

"They must be really small," Little Mike said.

"But they change when you push those buttons," I said, referring to the preset buttons on the radio.

"Yep, they are super tiny," Dad said. "If I turn this dial, I can get them to speak or sing in Spanish as well." Dad turned the tuning dial, and sure enough, his little band under the dashboard responded.

"That is so cool," I said.

"Most trucks and cars come with their own little band," Dad said.

Little Mike and I thought that was a great thing to add to a truck.

After a couple of years, and many times getting stuck in the mud later, Dad sold the Chevy and bought a 1969 Jeep Gladiator pickup. It was a J20, the Jeep version of a three-quarter-ton truck. It was awesome. Solid, heavy, dependable, and made for those difficult roads. Four-speed, manual transmission, four-wheel drive, and as tough as it looked.

Getting to and from the ranch now became a much more likely prospect, not to mention being fun if the road was wet and slick. Since most trucks in those days were single cabs, it was common for the kids to ride in the back. A lot of vehicles didn't even have seat belts, and if they did, the government was not involved with whether you decided to wear them or not.

That Jeep became the truck that Little Mike and I learned how to drive in. The lack of power steering coupled with our seven- and eight-year-old muscles meant that sharp turns weren't going to happen. This worked out great when the path was meant to be straight, and when we came to a tight bend, we had to take it slow since it was all we could do to turn the thing, anyway.

We sat on phone books to be able to see out the windshield. Phone books were thick in the days when they provided local advertising in the yellow pages. Just one was thick enough for me to get my long and lanky frame to a position where my eyes cleared the dashboard and I could still reach the pedals. Little Mike struggled more, and I remember having a toolbox behind him and a couple of phone books under him to be able to see over the dashboard, and he was leaning more than sitting in order to reach the pedals. He might have needed it that way until he was eighteen since he didn't have the same growth pattern as I did. Both of us had all we could do to operate the clutch and move the gear shift.

When either of us was driving, an exasperated and nervous dad was on the passenger side, and the child not driving sat on the tailgate with his feet dangling. Our boxer dog, Bandito, thought that dangling legs were the greatest thing to play with, and we would kick as he tried to bite. There were many times when the truck had to be stopped for a boot that had been pulled off, and a few times when the truck had to go back and get the kid that had been dragged off, still in the boot. Great times.

That truck went on to be the welding wagon, and I built pens with it hauling the welder and cutting torch. Then it became a trash truck at Pendaries Park, a resort property that my dad bought. When it had become less reliable, it went back to Dad's ranch and was left as a last-resort truck if a small job on the ranch had to be done. In 2020, I traded a McNab puppy to Dad for that truck and brought it back to Missouri. I had it fixed up enough to drive and painted the original Jeep green that it had been in its glory days. The Green Machine has become my daily driver in good weather. Something about driving it just makes me smile, even though I know I won't be getting there in a hurry.

7

WHITE WATER RAFTING

Below the big barn was a dry creek bed, except for those rare instances when it was not. Weather in the mountains of New Mexico can be such a splendid thing in the monsoon season. The sky takes on a hue where the colors of things appear both enhanced and dull at the same time. It is like things can't decide what color they really are as the muted sunlight plays with the tints and tones. Thunderheads can build in the mountains and the smell is one that has not been matched by man. The scorching, dry wind is replaced by cool, damp gusts, speckled with little droplets that tell a lie of things to come. When the big drops start to come, they are spaced far apart, like they had to maintain their personal space on the way down. Their fellows follow, and the result is often a deluge of Biblical proportions. Experienced mountain-dwelling New Mexicans will find shelter as soon as the first fat drop crashes into the ground. When hail is in the offering, it can be like being overcome by an ice wave on the edge of an ocean.

There were few things as satisfying as ducking into a barn or jumping in a truck to wait through these summer rainstorms and looking out as the ground darkens with wetness. The sound of the massive drops exploding on the metal can be deafening. When you couldn't get under something, you had to accept that you and your clothes were going to get soaked.

It was monsoon season at the ranch, and the dry creek below the big barn was no longer dry, or even a creek for that matter. It was a raging brown torrent – a rushing river that would do honor to any of its white-water cousins as the rainfall in the mountains upstream had been gathered into its embrace. Boulders and logs joined the water to give the liquid some solid substance and increase its damaging potential.

Dad had left us home with a typical list of chores: feed horses, chop wood, build fence, peel pine logs for the corral, fix the broken corrals logs, build new corrals, load adobe bricks, re-tin a barn, build a road, dig a basement, pour a concrete floor for a new shop, and frame in a new cabin after milling the wood from the trees on the side of the mountain. You know, a basic day that Dad would leave us to as he went to his office job in town.

"Whatever you do, don't go in the creek," Dad said as he left.

Those of you that have experienced life as a kid knows just how hard it would be to stay out of the creek with a command like that. Now, Dad didn't take any nonsense from us, so if we got wet, we would have to ensure we were dry by the time he got home. We had to make a raft, float the creek, and get our chores done before Dad got home. Our work was cut out for us.

As the Jeep disappeared over the cattle guard, I was rounding up the needed supplies to build a raft to take down the flooded creek.

"Let's go," I said to Little Mike. "We are going to have to hurry if we are going to get to do some white-water rafting."

"Dad said that we can't go to the creek today," Little Mike responded.

"I don't think he was serious," I said. "Hurry up, we have a lot to do today."

"Are you sure?"

Well, that was a silly question. Silly enough that I didn't bother answering him as I headed toward the big barn.

Five pine logs that had previously been smashed by wild cows in the corral and a few two-by-fours nailed on to hold them together, and we were ready to go. Little Mike later feigned innocence, but he helped drive the nails and drag our seaworthy creation to the banks of the torrent, making him an accomplice.

"Come on," I said. I was standing on the raft. My Huckleberry Finn-inspired pole was in my hand to shove us off.

"Are you sure?" Little Mike said.

There he was with that question again. I had never been surer of anything in my young life to that point, so my confidence convinced him.

"This is going to be great," I said. "We'll ride to the water gap below the big field and maybe drag the raft home."

The fact that I had formulated such a well-thought-out plan inspired Little Mike's trust and he stepped onto the raft as if it were the thinnest of ice over a pond filled with gators. Although tentatively, at least he was there. The current caught us, and we were in for a fantastic float trip. Within seconds the raft was starting to show some signs of its engineering design flaws. Research and development were telling me that adding some baling wire to this design in support of the nails would have been beneficial, and I was making mental notes as we went. We had been swept downstream three hundred feet or so when the biggest shift in our platform occurred. There was at least a mile of fun ahead of us, so the problem was a substantial one. The logs were still parallel but no longer even with each other as the rectangle became a parallelogram and eventually lost all shape entirely. The raft disintegrated in the raging water and rolling rocks. Those boulders were sharing a similar experience with Little Mike and me.

By the time I found the bank again, we had achieved more distance without the raft than we did with it. I was on the other side of the creek, and Little Mike was still with me. It was a great time, but there was going to be the problem of getting back across the water in order to get home and get some work done before Dad made his impending appearance.

With the aid of broken trees and a washed-away fence that served as a makeshift support, we made it back to the side of the creek that the house was on. Little Mike was a little worse for wear, and his trust in me had dwindled to one of its lowest points. Because of the obvious colorful bruises and other harms, hiding the raft trip was not going to be the easiest thing, whether our clothes were dry or not. In fact, it proved impossible.

The little barn in those days was more for storage and tools, while the big barn was where the tack rooms, hay, feed rooms, stalls, etc., were. Feed still came in brown burlap sacks in those days, and they had the value of a nickel each when returned to the feed store, much like a pop bottle did when those were still made of glass. We had great piles of the sacks bound together by the ever-present baling wire. These piles were of a convenient height to bend a youngster over when spankings were to be administered, so the little barn was, to Little Mike and me, the proverbial woodshed.

That particular spanking was more robust than some in my memory. It was fueled by Dad's fear of our death and his subsequent joy at finding us still living. I had to move on from this experience and use future designs of the raft only in the stock pond, which we did when my backside had recovered from the belt and my other anatomy healed from the epic whitewater rafting adventure.

8

PEN FULL OF COLTS

We had a neighbor named RD, one of those guys who was always trading, collecting, bargaining for, and moving things, animals, and people. His was the closest place to us, probably a little less than two miles away. Every time we went to his place, there was a new group of people. Some were in trailers, some in tents, some looked fairly normal, and then those who looked like the carnival had rejected them. His horse remuda was similarly changing all the time with horses to match the people. I guess you could call RD a collector, although one without much in the way of direction.

"I've got a bunch of young horses for you, Mike," RD said to my dad.

They were leaning on the hitching rail at RD's place. I seem to remember that RD was always eating. The mayonnaise had squished out of the bottom of RD's sandwich but was saved from the ground by his ample belly, joining the multitude of stains competing for prominence. My dad turned his attention to tamping the tobacco down in his pipe and didn't immediately answer.

"They'd be perfect for them boys to start learning how to train," RD continued.

"I think the boys are a bit young for that," my dad said.

"Nonsense. Young boys need young horses," RD slapped my dad on the back. "I'll bring them over and drop them in the pen at the big barn of yours and you can see what you think. No commitment now, you understand."

"That's OK," my dad said. "We aren't needing them."

"OK. I'll bring them over tomorrow," RD said, squirting more mayonnaise out of the bottom of his sandwich with his teeth, smiling through the dangling ham and cheese.

Getting the five, two-year-old horses wasn't really a hardship for Dad, but he was right. We didn't need them and certainly weren't ready to start working with them. I had to be around eight or so at this time, so Little Mike was six or seven.

RD brought the horses over as promised and put them in the large corral that was part of the big barn. This barn was a very cool structure, measuring somewhere around fifty feet wide and eighty feet long. It was built on the side of a hill in such a way that the side that you could drive in on was level, but the opposite side was open and about four feet higher than the ground. There was a long trough on the downhill open side that you could sweep hay into directly from the concrete floor of the big barn, but the height of the floor was tall enough that animals could not get into the barn from that side. The trough side made up one edge of the big corral. It was a well-designed system that I always liked.

"There's some new horses in the corral that you boys can look at, but I don't want you to mess with them," Dad informed Little Mike, Rick, Gus, and me as he headed out to work that day. "I'm serious now. Do not touch them."

Of course, we raced down to have a look at our potential mounts.

"I like that red one," Gus said.

"Why don't you try to catch him?" I suggested.

"OK. Come help me," Gus said.

Grabbing ropes, we all jumped down into the trough from the floor of the big barn. As we climbed out of the trough and into the pen, the skittish horses ran to the far end of the corral in a group. It was the four of us cousins against the five wild horses.

We moved, they moved. Our strategy was not planned nor educated, and there were several near misses as the horses almost ran over one of us trying to escape from someone else. It was mayhem. Fun mayhem, but still mayhem.

"Why don't we use some grain?" Little Mike said.

It was a brilliant suggestion, and I jumped back up in the big barn to get some.

I scooped the grain out of the sack with a coffee can and dropped some in the trough. The tamest of the young horses came right over with the promise of sweet feed.

"I think I can get on him," I said.

I had the height of the big barn floor on my side and a horse that was interested enough in the grain that he was standing still for a second. I jumped from the trough and landed on the horse. He immediately jumped, and I landed on the ground.

"That was great," I said. "Let's do it again."

The next several hours were spent getting horses close enough to the trough or a fence to get on them and consequently landing on the ground time and time again. By lunch, the horses had at least all been touched and each of us had all had taken several chances at trying to ride.

Of the five, there was a spotted one that was willing to let us sit on it without running off or bucking, and we each ended up getting a chance to sit on it.

"What have you boys been doing?" Mom asked when we went back to the ranch house for lunch.

"Just looking at the horses," I said.

"You guys look like you have been doing a lot more than looking," Mom said.

"You know Auntie Diana," Gus said. "Just trying to catch them."

"Well, you boys be careful," my mom said. "Lunch is ready."

Mom's lack of concern just meant that she didn't fully understand what we were up to. Her domain was in the spotless house that was always full of good food and comfort. If she had had any idea of the danger that we were creating and facing daily, I am sure she would have insisted on moving to town. Dad knew what could happen, he just wasn't fully aware of what was happening all the time and wasn't about to let Mom in on it and potentially lose his country living.

We spent the afternoon working on whatever chores had been left to us that day. I don't really remember what they were because our minds and thoughts were on the pen full of horses.

"How was your day?" Dad asked when he got home from the office in town.

"Great," I answered.

"Did you boys look at those horses?"

"Oh yeah, they're great," Little Mike said.

"Which one did you like?" Dad asked.

"I like the spotted one," I said. "It was the best one to ride even though it was scared to come up for grain at first."

"What do you mean ride?" Dad asked. An edge had come into his voice.

None of us said anything because we just recognized the error of my comment. It was easy to tell when Dad was mad, and I was very good at getting that look from him.

"Did you boys touch those horses?" Dad asked. His voice was quiet, which was more frightening than if he had been yelling.

We exchanged looks, very much as you can imagine, with Adam's apples bobbing and eyes darting.

"I think we best go down to the little barn," Dad said.

Well... You know the rest of this story.

9

SNAKESKIN FORTUNE

"Hear that boys?" Dad asked, pulling up his horse.

"What?" I asked.

"Just listen. There is a strong buzzing sound. It is from a rattlesnake that must be under that rock we just rode over. Come on over here."

Dad rode toward some trees and got off his horse, handing me the reins. "Stay here for a second and I'll see if I can find it."

He walked back towards a large flat rock that we had just ridden past. Little Mike and I were sitting on our horses in the shade under a dark-green piñon tree. We could hear the rattle from where we sat, but none of our horses seemed even the least bit bothered by it. I guess they lived their lives amongst that sound on enough occasions that it didn't bother them. Hollywood would have you think that every horse bucked their riders off in just the right location for a rattlesnake to bite them when one was close.

"Tie them horses up and come over here," Dad said.

We dismounted and tied the horses to the piñon trees.

"He's right under there boys. Don't get too close."

Dad had found a long stick and was pointing at the snake that was now coiled and shaking its rattle with reckless abandon. When the stick got close, it would strike at it, trying to bite the threat.

The brownish-tan-black snake with a suggestion of green slithered out toward dad, and he pinned the massive head to the ground with the stick. Picking up a fist-sized rock, he smashed the head and killed the snake. With the dead snake's head still pinned, Dad took his pocket knife out and severed the head.

"You can still get bit when they are dead, so you've got to cut the head off. Let me show you boys how to skin a snake."

He lifted the headless body and slit about three or four inches of the skin along the yellow-scaled belly leading up to where the head had been cut off. Peeling it back like you would a tight sock on a sweaty leg, he pulled the body of the snake out of its skin. The white and bony meat came out and the skin was turned inside-out in the process until dad got to the tail. He had to cut off the last bit of the snake to keep the rattle still attached to the skin. Working it back so that it was no longer inside out, he had the long flat skin of the snake with a rattle on the end.

"That's so cool," Little Mike said. "What are you going to do with it?"

"Well, you can do a lot of stuff. How about we make a snakeskin belt?"

"That sounds great," I had to agree. "How will you do that?"

"We'll just put your belt through it when we get home and let it cure."

Dad rolled up the skin and we walked back to the horses. There was a lot more skin length than the measly bit that I required for a belt, but dad put one of my leather belts through it and I ended up wearing a rattlesnake belt around my waist for quite a long time. Little Mike ended up with a snakeskin belt as well, so we were a fashionable pair. Rattlesnakes were not in short supply.

One day I came across a nest of about five or six tiny rattlesnakes. Picking them up, I placed them in a five-gallon plastic bucket.

"Where did you get those?" Little Mike asked, looking into the yellow bucket that I was carrying with the unfortunate baby snakes slithering around.

"Up by Wayne's cabin," I answered. "They're baby rattlesnakes. See the little button on the end of them?"

Pony, an old man that worked on the ranch, had a brother that my dad had let build a little cabin on the side of the mountain. I remember it as a very neat little place until Wayne burned it down. Drinking, smoking, and the super dry wood of a New Mexican-climate cabin can be a flammable mix. The main thing I remember about Wayne was his scraggly, smoke-stained, white beard and his missing thumb. And he built a pretty cool cabin. At least from the perspective of a seven-year-old.

"Those are the smallest rattlesnakes I've ever seen," Little Mike said.

"Yeah, they're just babies, so they can't hurt you like a big one," I said, grabbing one and showing it to my brother. Displaying my superior knowledge of all things snaky.

"Good thing," Little Mike said. "I wouldn't be handling them if I were you just in case. What are you going to do with them?"

"Don't worry. I put all of them in the bucket without a problem. I think we should make hatbands. We could make a fortune. Everyone would want a rattlesnake hatband."

"Yeah," Little Mike agreed. "That's a great idea. They just need to get a little bigger I think."

When Little Mike was right, he was right.

"Let's find a place in the Big Barn to keep them while we feed them up. What do you reckon they'll eat?"

"Maybe some of those bats that we keep shooting in the rafters."

"Let's put some bats in there and find out. But don't sound too tasty to me," I said.

We walked back to the barn to find a good place to corral our future fortune until they got big enough to make it around a hat.

"What you got there boys," Dad asked as we walked up to him. He was fixing another broken log on the constantly broken corral.

"Chris found a bunch of little rattlesnakes," Little Mike said. "We're going to make them into hatbands like our belts."

I walked up to dad with my yellow bucket. "Here I said," reaching toward one of the snakes to hand it to him.

"WHOA. WHAT ARE YOU DOING!?!?!" Dad yelled.

He was obviously scared of snakes, so I tried to calm his fears.

"Don't worry Dad," I said. "They are too young to bite you."

"Do not put your hand in that bucket son," Dad said, leaving no room for conversation. "Baby snakes are just as, if not more than, dangerous as a big one. Their venom is highly toxic. What on earth were you doing handling those things."

He shook his head and I'm sure he had to wonder whether or not I was simple. This was not the first time that my dad doubted that I would live to be an adult.

"I'm going to kill these ones before they get big," Dad said. "You boys leave all snakes alone. Period. You hear me?"

Dad took the bucket and walked off toward the tack room. He was muttering to himself under his breath and Little Mike and I knew that we were not invited to follow. There was the noise of a shovel smacking the ground, the snakes between the metal and the ground. We didn't end up with any skins that we could have used to increase our wealth. I tried to ask Dad about that, and he only repeated that we were to leave all snakes alone.

I was still at that precious age where God was still looking after me. Young enough to blame it on my innocence instead of my foolishness.

10

BUG BATTLES

"Well Dr. Bush, he has been complaining about his ear hurting, but I can't see anything in it," my mom explained to the doctor that was about to start examining me. "It's been swollen like that since yesterday.

"Does it hurt when you swallow?" Dr. Bush asked, pressing on my neck. "How about here?" he asked, pushing a little on my jaw.

"Just this one ear," I said.

"It does look red and swollen," Dr. Bush had to admit. "Let me see."

He took the little tool shaped like a funnel with a long handle and a light and looked into my ear. "Mmmmmm hmmm. I do see something. Hold on now," and he grabbed a pair of tweezers to go after the object.

"What is this?" he asked, pulling a wax-covered ball out of my ear.

It was ant shaped, with a couple of dead legs sticking out of the ear wax that had imprisoned it.

"I think this is an insect," Dr. Bush said. His powers of observation were keen. "I believe this is a big red ant."

He was correct, and that big red ant had caused me some serious misery. This was not the first red ant to sting me, and not even the worst place I had been stung. Mom and Aunt JoAnn had seen me through the windows of the ranch house as I stripped with the speed of a whirling dervish, putting all my anatomy on display in an effort to rid my nether regions of the feisty little creatures. They couldn't tell the story all the way through for their laughing. I would take getting stung in the ear instead of the even more sensitive bits if given a choice. The ants were indiscriminate in their stinging and whatever part they chose became a 'sensitive bit'.

The ranch was covered with red-ant hills. These were small symmetrical mounds of light-colored sand that were the work of thousands of red ants. Some of the larger ones would be at least six feet across and stick out of the pasture two or three feet in elevation. These ants were aggressive and defended their territory with zeal. They provided a lot of entertainment value when you put a large bug amongst them, but that entertainment might come at the price of a few stings. The ants would attack the bug like "ants on a grasshopper", which has become a favorite saying of mine over the years.

My favorite victim to bring the ants were the large black stink bugs that were common on the ranch as well. When they sensed a person close, these big beetles would stop and point their backsides straight up in the air, letting out a noxious scent. Since I was endowed with such an inquisitive nature, I had picked one up and was trying to see where the stink was coming from. That beetle let out a stream of brown liquid, firing it right into my eye. That was painful and educational, as many of those types of things are, and I had decided that stink bugs would make great ant food from then on out.

But now that the red ants had caused me to endure an actual doctor visit, it was time to do something about them.

"Want to come along and help me find a red-ant queen," I asked Little Mike as I was gathering up shovels and buckets for my search.

"No way," he said. "Those things sting, bad."

"Not if you can get them off in time," I said.

"You can't always do that. Just ask Aunt JoAnn and Mom about your naked dancing," he laughed, referring to their story. "I think I'm good. You go get the queen."

With that, I was off again to get the queen and kill the red-ant colony that I was pretty sure had sent their soldier into my ear. It had been a big one just on the other side of the creek below the big barn.

The creek had cut into the ground, leaving about a four-foot-high ridge of New Mexico clay and a deeper hole in the bottom of the creek that was a favorite area to play in for Little Mike and me. Not deep enough to swim in most of the time, but a great place to have fun. The little ridge around one side of it was dotted with small round holes that we always expected a snake to be peeking out of, but there never was. My designated ant hill was about ten feet beyond the little cliff. We had been at the creek a few days before and I had been feeding the red ants whatever hapless bug I had been able to find. This had been their opportunity to assault the depths of my ear.

I made it across the creek with my shovel. It had occurred to me to tape my pants around my boots so that I could have an ant-proof uniform to go about my work. I started digging and brushing off ants as they sent their soldiers out in force. Their great battalions turned the ground from tan to red as they came to stop the storming of their castle. I would get a shovel full of sand and ants and send them flying through the air as far from where I was digging as I could.

The hole in the mound was getting deeper and I had only suffered a couple of stings. These were on my arms, so the slight pain was endurable. Another one got me on my back. Hard to get to that little bugger, but he paid the ultimate price. Then another on my neck. Now they were trying to get back in my ear. Then another on the underside of my arm. That was a bit unfair. I felt one on my leg. How he got in there was an impossibility I was trying to work through when another one got me on my leg. It was getting too much and I could feel dozens crawling in my pants. Then a brave ant was able to assault my most delicate of places, and my quest was over.

The truth of it was that there were probably only a few ants in my pants, but the mind does what it has to so that the body is pressed to action, and I fled the scene, trying to get my clothes off as fast as I could as I ran toward the cliff above the creek.

Taping my pants to my boots had seemed such a good idea until I needed those pants off in a hurry, and I stumbled and fell in my rush to get away from those soldiers and out of my clothes. I was far enough from the house that they could not see everything unfold in intimate detail but close enough that they could see the gist of it. Pretty much the distance from the top of the stadium to the middle of the field at a pro football game. Still close enough that I could hear the laughter drifting from the yard at the ranch house as Little Mike gathered my folks and grandparents to watch the comedy unfolding in front of them. To be fair, it had to be funny enough to get even the most somber to let loose with a belly laugh.

I finally got my pants off and I jumped naked over the little cliff and into the water. I was brushing and pulling ants from several places, the water providing a bit of cool to the inflammation that the ants had caused. When I finally made it back to the house in just my boots and whitey-tighties, I looked like I was suffering chicken pox or some other malady that had left me spotted with red welts. You would have thought there was a limit to how much people could laugh, but my family was pushing the limit to the max and couldn't help themselves and I made my way into the bathroom to take a shower and clean off the dead ants and mud that clung to my form.

I eventually dug up a few queens over the next few years. The experience gained from the first attempt advanced my attention to personal protection, and I spent as much time taping up openings in my clothes as I did in the actual digging. Those red ants and I remained enemies as long as I lived on that ranch.

11

IN THE MOUNTAINS

We spent a lot of time in the mountains on horses. Some of the most vivid memories of my childhood include riding in the mountains. One of our first trips was when I had to be around five or six, so Little Mike would be between four and five. We ended up getting caught in a mountain rainstorm, soaked through, cold, miserable, and letting our emotions rule the day. Dad had to be wondering about including his children in the fun at that point, but he persisted, and we went on a lot of rides through the wilderness.

Looking back on it, I have come to realize that I was the more expendable of my dad's two offspring. Not that anyone who knew Little Mike and me at that age would blame him if something happened.

On one of the trips that I remember, we were riding on a trail with a place that was decidedly dangerous for man and horse. The trail came to a ledge that stuck out from the mountain with barely enough jutting stone for an iron-shod hoof to fit. A rocky death was promised to any person or beast that fell. In my memory, the ledge looks longer and thinner than it probably was, but it certainly was dangerous enough to hasten that unavoidable meeting with your maker. Legend has it that this place had been the end for several unfortunate fellows and their horses in the past.

We had been going along for some time, and I was on a horse that was notoriously stubborn when he didn't want to go but not dangerous in the way that he would buck or run off. Little Mike was on a different kind of horse that was all go, but might buck a little, leaving you in its wake. It was the type of horse that tossed its head and wanted to stamp and prance when the group stopped and was generally not fun to ride. His horse was the type that would turn sideways to the trail at odd times, or would drop its head to eat when grass was spotted. If you have ever ridden the type, you can imagine this horse's attitude and probably even see the horse's facial expression.

As the trail brought us closer to this slender ledge, Dad stopped and said to Little Mike and me riding behind him, "Boys, I think you should change horses to go across this next little bit."

"No problem," I said, getting off my mount and handing him over to Little Mike, never questioning or wondering why Dad would want us to do that.

Little Mike got off and we exchanged horses.

One hundred yards past the widow-maker part of the trail, Dad pulled up again. Letting out a long breath that he must have been holding for the duration, he said, "Well done, boys! You can trade back now." He was obviously relieved to be spared having to explain to Mom how he had lost one of their children.

Little Mike and I traded back to our original steeds and the group carried on. Not much was said about it, and no one seemed interested in why we changed horses. It wasn't until later years that I reflected and wondered about Dad making that switch. You may remember that I indicated I was a difficult child. Little Mike was easy-peasy compared to me, so I think Dad thought it might be easier if he had to explain to his wife that he lost a kid but at least it had been the trying one.

Another time we were going through a valley in the Pecos Wilderness that had belly-high, dark green grass and was crisscrossed with small streams that could not be seen for the grass. In my memory, it was called Burro Valley, but that might be another valley where the hail caught us. Anyway, I was on a little bay horse that I remember as Fargo. Fargo hit the edge of this valley and dropped his head to the grass. He was easily able to deny the strength of my seven-year-old muscles, and he was not about to miss out on this oasis of beautiful green just because of the skinny little irritant on his back.

Dad was leading one of the pack animals, and he rode up behind Fargo and swatted him on the rump with the end of the lead rope. Hard. Well, Fargo and I hadn't seen that coming, and he quit the earth. His first jump would have cleared a bear if it had been in front of him, and he was moving like he thought there was one behind him. We were screaming across the valley like a bullet, and thankfully Fargo was able to see the streams, even if I couldn't. Maybe he sensed them, I don't know for sure. He would leave the ground about ten feet on one side of a stream and land about twenty feet on the other side. It would have been inspiring to watch, but it was not so inspiring to endure.

One of the two main things in my favor was the saddle I was riding. It was an old Frazer with huge swells and a tall cantle. It was built for a kid my size, but also built so that a kid my size would almost have to have the horse turned upside down in order to come out of it. Riding in it felt just short of being wire-tied to the saddle (experience talking there). The other thing in my favor was that this was not the first time I had been on a runaway horse. By the time we made it to the end of that valley, I was enjoying the ride and Fargo had decided that there really wasn't a bear behind him. When Dad and the rest caught up to us, Fargo had his head buried in the grass again and I was pulling on the reins, thumping with my feet, and trying to move him along the trail. Dad swatted him again, but not as hard, and he let me know he was about to do it so that I had some control when Fargo's head came up.

The hail story was again crossing a valley. I was a bit older by then, so my ten-year-old strength and experience were a match for any horse that wanted to graze. We were about halfway through this valley when one of the pack saddles began to slip. Dad and Pony stopped to fix it.

Thunderheads that were gray giving way to black had built up along the mountains ahead of us, and we were pretty sure we were going to get wet. The wind, humidity, and temperature all changed in a way that makes the hair on the back of your neck stand up. Dad and Pony were working like squirrels on a birdfeeder when the solid white wall of hail made it over the ridge and advanced on us in that valley. We were just like proverbial sitting ducks.

It was a wall that looked like it had structure and strength, its architect having a vision for the immense. This wall was moving toward us at an alarming clip, and we started for the tree line. Dragging panniers and horses behind us, we were too slow to keep the hail from pelting us. Thankfully, the pieces of ice were the size of peas, so bodily damage was not likely. Their diminutive size didn't keep them from being cold though, and the warm day had dropped from somewhere in the 70s to somewhere in the 40s in a matter of minutes. We shook hail from our clothes and brushed it off the horses as we sat there in the trees, happy for the poor shelter that they provided. At least it was something.

When the storm had passed, it left about four inches of ice on the ground and looked as if it had snowed. It was an awesome display of God's power and influence. The weather has made men seek their faith since the beginning of time, and seeing this happen in front of you would certainly have that effect. I remember it like it happened yesterday.

12

PONY THE COWBOY

A tall and wiry old cowboy named Pony came to live on the ranch at one point. He had been another of our neighbor RD's collectible humans but was lovable enough to pass on to a good neighbor like my dad.

Pony had a past that deserves a book and movie of its own if half of his stories were true. He had run away from home when he was twelve and spent a lot of his youth rustling cattle in Mexico and bringing them to the US. He had roped a mountain lion, been shot by a Mexican firing squad for stealing cattle, shod horses for the cavalry, had nearly every bone in his body broken at some point, outlived three wives, and had other significant adventures. He smoked Bull Durham roll-your-own cigarettes, put Jim Beam in every cup of coffee, and could do just about anything with a horse.

Of course, a character like that coming into your life as a kid was amazing. He lived in an old Airstream camper that was plugged into the big barn. I don't know what arrangements Dad had with Pony, but I feel like he was expensive; maybe not in direct dollars, but in the horses he killed and the equipment he ruined. Sometimes cheap labor can be the costliest.

Those ancient homes of San Pablo that we had been dismantling for adobe bricks came in handy when it came to housing workers on the ranch. Luckily, we hadn't needed every one of the ruins, so there were still some standing, if the word "standing" could be applied. While many might consider the house we lived in as a ruin in its own right, it was a palace compared to the remains of San Pablo. After Dad finished with his remodel and Mom did some interior work, that home could be considered quite the mansion.

With Pony now on the ranch to help get more of the work done, he needed to find a few more workers. Dad left it up to Pony to find his men, and Pony found some hippies, who were more commonly available then than now.

There were a few transient folks who occupied the ruins through the years, but the two I remember best were Pete and Claude. Pete had come from some eastern city and had the dream of living off the land in the mountains of the southwest. He had a girlfriend named Betty, and they were the model of hippie happiness. Homemade leather leggings for Pete and flour-sack dresses with flowers in her hair for Betty. They lived in one of the ruins, and Dad allowed them to have their herd of goats as part of the perks of working for Pony on the ranch. As I remember it, Pete made $5 a day to thin trees on the mountain. There is a good chance Dad didn't get his money's worth out of Pete, either.

Claude was another hippie that lived in one of the ruins. I don't remember him ever working, but I do remember him being around. I had never met a man named Claude before, so I was confused about his name. My first recollection of Claude is of Pony telling my dad about him.

"Come on in," Pony said as Little Mike, Dad, and I as we entered the Airstream.

Oh, those mornings in Pony's trailer. We would sit at the tiny table. Dad or Pony would put a capful of Jim Beam in our coffee, and no one minded if Little Mike and I tried to roll a cigarette and coughed through taking a puff or two. The communal spittoon sat on the table like a centerpiece.

"I've got another'un named Claude," Pony said. "He's jus' 'bout fixed up from a mountain lion attack."

"No kidding," Dad said. "What happened?"

"Story is, he shot a turkey an' was carryin' that turkey on his shoulder, walkin' through an arroyo," Pony said. "A big cat must've stalked'im an' jumped on that turkey. All but scalped ol' Claude, not that he had any hair to start with 'cept the ponytail in the back."

"Did the mountain lion maul him badly?" Dad asked.

"Nope," Pony said. "Three big tracks cut from 'is forehead across the top of 'is skull is all. The cat took the turkey and left ol' Claude layin' in the bottom of the draw like he'd been left by injuns. Pete said he lost enough blood he's loopy fer a bit. I reckon it might've been a 'cun-cushin.' Chances is he's a loopy to start with anyhow, don't know as I ain't spent much time with'im yet."

"Wow," Dad said. "I guess he's pretty lucky he had the turkey for the mountain lion to take."

"Yup," Pony said. "I reckon that cat thought a turkey was ridin' a goat down the draw. Claude said his neck's still sore, but the wounds are gonna leave scars like the ones on my left butt cheek."

Pony had told a story of roping a mountain lion off his horse. It had run up the rope, attacked him and the horse, and he had scars on his butt from the lion raking his claws across it. I don't remember seeing those scars, but I do remember the story, and I remember people saying they had seen the claw marks.

"Is that why they call him Claw?" I asked.

"His name's Claude," Pony said.

"I know," I answered. "Is it because of the claw marks?" I still hadn't understood the name.

"Sorry, son," Dad said. "It's Claude with a 'd' at the end. It's just a name."

"I think Claw would suit him better," Little Mike said. I had to agree.

As time went on, Betty moved in with Pony. He was in his sixties, and she had to have been in her twenties. He told Little Mike and me that he needed a cook, and I can see now how a man would want that. We were too little to understand. Pete was not agreeable to Betty's new duties and living arrangements, so he moved on. Dad had to kick Pony off the place as well in a story I won't relate, and the remaining hippies found easier climates to live in. After that experience, Dad mainly relied on day help. Old cowboys and hippies were just too expensive in the long run.

13

STREET SMARTS

Aunt JoAnn was living her big-city dreams. Her airline stewardess occupation had led to some incredible adventures, and she was living out what most young girls of her era could only imagine.

She had married Chicago's most eligible bachelor, and they ended up having two amazing kids, Christine and Jeff. Nicer and more generous people you will never meet, and they have been a delight in the role of first cousins throughout my life.

Since they lived in Chicago, the visits to the ranch were fewer than they should have been. My influence on their young lives was restricted to a week or so a year, but Little Mike and I did teach them many life skills such as turning a corner just right so that we could practice throwing a heel loop at their feet, and how to carry your body weight in stove wood when you filled up the wood box. They remember well the hours they spent being roping dummies for us and worker ants for Dad.

Being family and not visitors, there was no escape from the work when they came to visit the ranch. Dad would gladly work anyone, willing or not, and Jeffery was more willing than most. To him, it was a grand adventure that he knew had an expiration date. Besides, it made his hero status grow amongst his city-dweller friends.

We were up early one Saturday on the side of the mountain, cutting wood for the stove. This was a constant job since Mom cooked every day on the wood stove.

"Uncle Mike, I have to pee," Jeffery said to my dad.

"That's fine son," my dad answered. "Just step over there in front of the truck and pee."

Jeffery followed orders fine, so he went in front of the truck to relieve himself. He walked back to where we were splitting and stacking, the front of his pants now wet.

"What happened?" I asked. "Did you pee yourself?"

"No, I just did what Uncle Mike told me to."

"He told you to pee yourself?" Little Mike asked.

"He just told me to go in front of the truck and pee," Jeffery answered.

"You're supposed to unzip and pee outside," my dad explained.

Little Mike and I were laughing, and Jeffery took it in stride. Sometimes lessons can be hard to learn, but he learned that one well. In fact, he learned it so well that he became a teacher, and he went back to the Chicago suburbs and taught all his friends how to pee in their local park. Much to Aunt JoAnn's displeasure, the gang of boys marked their territory all over their part of what passes for the woods in Chicago.

14

PUBLIC SCHOOL

Middle school was uneventful as far as I can remember. It was not academically challenging, which had been the source of many of my problems up to that point. Mom had taught me to read well before kindergarten, and I had taken to it with a passion. This had rendered most of school quite boring and mainly a place to goof off and get in trouble. In the modern world, there is no doubt my teachers would be clamoring for drugs to curb and direct my personality.

There were a few progressive teachers, and the system allowed them to think outside the box. Mrs. Sanchez, the seventh-grade English teacher, allowed me to go to the school library instead of attending class. The deal was that I had to read an assigned book every week, write a report on it and turn in the report on Friday. There was no way to finish the book in the hour allotted each day for English class, so I ended up needing to read a lot after hours to stay caught up. Even though I did not always like the book assigned, it laid a lot of groundwork that I was able to use to get through college more easily. I was the only one in the class who got to do this, and I don't really know for sure if Mrs. Sanchez was trying to help me or just get me out of the classroom so I wasn't a disturbance to the other kids.

Either way, I felt special. Maybe window-licker special, but special is special.

As you can tell from reading my work, I did not learn how to always construct the perfect sentence or avoid the use of run-on sentences because I was in the library instead of my school desk during the time of that important instruction, but I did continue and build my love of books.

I made friends with the janitors and got to hang out in their little area as well. It was like hanging out in the engine room of a big ship. It was the warmest place in the winter with the water heaters and boilers in there, and it was a great place to hang out and chew tobacco when there was snow on the ground. Interestingly, it was also the grubbiest and dirtiest place in the building, despite being the area where the cleaning people hung out.

"Give me a dip," Tony, the janitor, said. "I like Happy Days more than Copenhagen, but I'm out."

During that era in America, it was common for the ranch kids I knew to chew or dip. We would have our bags of leaf, dried tobacco plugs, or snuff cans, and no one seemed to mind. If you can imagine, there was a time in America when you got to make your own decision about what you did, whether it was safe or not. You could even ride around without a seatbelt. I'm not condoning the practice, but I do think free men should be free. If my decisions are not hurting you, I should be able to make them at my own peril. Sorry, back to the story.

I passed my can to him and said, "My dad does Happy Days sometimes. You should have seen my cousin puke from Skoal."

"Skoal's the best," Manny, another janitor, said.

He passed me the coffee can that was our communal spittoon. I kept it at arm's distance because the smell of Skoal was offensive to a Copenhagen connoisseur such as myself. The truth of the matter though is that an experienced spittoon is offensive in just about every way, Skoal only adding another layer.

"I have some Day's Work," I offered.

"That stuff will stay with you," the other janitor, Jack, said. "One bite of that plug and you've got flavor 'til supper."

"You get paddled again?" Manny asked me.

"Yeah," I said. "Mr. Armijo's got it in for me."

They laughed.

"Maybe you throwin' snowballs might have had more to do with it than a principal with a grudge," Manny said.

"Only Tim and I got paddled though," I said. "Eric and Steve were throwing them, too."

"You got to be smart about it and stop throwing when the teacher shows up," Jack laughed. "You won't get paddled if you don't get caught."

Everyone nodded their agreement at this wisdom.

Those men were always very nice to me, and I don't remember any other kids getting to hang out in there with us. There wouldn't have been space, anyway. That was another special situation I had found in the routine of things. This has been, and continues to be, my pattern.

15

ALLOWANCE

I had many friends in town who were experiencing a childhood that I much envied. We had not had television since I was about seven years old. Dad had walked into the living room and asked my brother and me how we were doing. Our attention had been glued to the adventures of some cartoon character, and I had answered my dad by asking him to move out of the way of the TV. The TV disappeared that day, and I did not have one for over ten years after that, at which point I had already been out of the house for two years. I don't remember anything about that cartoon character, but I can still see my dad's face.

My friends in town got to go be kids on the weekends. Bowling alley, bike rides, sleeping in, trips to the park, etc. For Little Mike and me, weekends generally meant chopping stovewood, digging post holes, or riding great distances to check on cows. I was allowed the occasional sleepover. Dad saw this as one more youngster to involve in some activity involving sweat, and my friends were not accustomed to the rigor. Not once did any of them accept a repeat invitation.

I had one friend who was getting an allowance of $5 a week. This seemed such an exorbitant amount to me in 1980 when I was all of twelve years old. I could not imagine what great things could be purchased for $5 a week, but I was willing to find out.

"Hey Dad, I have a lot of friends who get money from their folks every week. It's called an allowance."

"What do they get that for?" my dad asked through his pipe smoke.

"You know, normal chores and stuff like cleaning their room and taking out the trash," I answered.

"Well, do you think you'd like an allowance?"

"Yes sir," I answered at once. "I think that would be swell."

"What kind of money were you thinking?" my dad asked. He had set his pipe down in the ashtray since this was now a business meeting that needed his full attention.

"Well, I have one friend that gets $5. I was thinking that $3 would be kind of neat."

"That just doesn't seem right to me after all the work you do around here," my dad said. "What do you think about $10?"

Had I heard that right? I could not fathom such wealth, but I eagerly agreed with my dad, having found him in such a generous state.

"OK," he said. "We'll make your first allowance on Friday."

To say that conversation had gone way better than expected would be quite an understatement, and I went about the rest of my week with newfound enthusiasm and determination in my appointed chores.

Mom always made a full breakfast. She was a wonderful cook on that old wood stove, and the morning was always full of amazing food and smells. Of course, it was even better on that Friday morning as I sat down to a plate of six eggs, six pieces of bacon, two pieces of toast, orange juice, coffee, and water. My bony frame made a lie of my appetite, for which I was well known, and this breakfast was pretty much my normal intake.

In front of my plate was an envelope. It was sealed with my name printed neatly on the front. I tore it open to receive my first allowance and found a crisp $10 bill that Dad had taken the trouble to go to the bank and get, brand-spanking new. Crisp, never folded or bent, it smelled of the mint it had come from without the grime of circulation.

Along with this glorious legal tender, there was an itemized receipt – or in this case, an invoice – that was unpaid. Dad had kept track of my meals, large as they were, added sums for laundry service, housing, transportation, and other stuff that adults pay for and kids enjoy. The total at the bottom was just north of $16. He was giving me wholesale prices instead of retail, as he pointed out, seeing how I was family.

"Well son, I guess I'll take that $10 back and you can get the rest out of your piggy bank," Dad said. "The lucky part for you is that we didn't start the agreement until Tuesday, so your bill only reflects a few days. I'd expect next weeks to be a lot higher, the way you eat."

"What do you mean?" I asked. I was astonished at how fast this allowance was turning out to be a bad idea.

"Once you start earning a living, it's time to start paying for things. Now that you're getting paid for doing stuff around here, I reckon it's time to start pulling your weight financially as well. You'll notice that you are getting a very good deal on food and housing, as I mentioned. Benefits of being part of the family."

It was a devastating wake-up call as Dad took the $10 and I reluctantly went and retrieved the balance to settle my bill.

"Do you want to keep getting an allowance?" Dad asked.

"No sir," I answered. "I think we should go back to how it was before."

"That sounds like a good idea to me, too," Dad said with a smile.

Lesson learned. A good one at that.

16

CHOLLA AND SAMBO'S

Since we did not have a television, the main way that Little Mike and I were able to get a TV fix was at Cholla's house. There was a tradition of eating tacos at Cholla's every Wednesday. She would make tapioca, white cake with chocolate icing, and tacos for all of us. I remember those meals with great affection. As everyone knows, no one cooks as well as your grandmother.

I had an appetite that none in my family could match. Little Mike will gladly tell of my eating in those days. As he spins the yarn, he will pose with his finger pointed and a sad/puzzled expression on his face as he does an impression of my dad watching me consume my massive caloric need. Mom and Cholla would make the tacos one at a time, so you would have another arriving as you finished the one on your plate. I would generally eat six to nine tacos, depending on how much cake and tapioca I was going to be chasing it with. Little Mike and I did not like the red chili in the meat that she would add to Dad's tacos, so we would eat ahead of the rest of the family. The red chili came from the dried pods that were always hanging in a bundle on a hook on the back porch. She would take them down and blend them before putting them in meat or enchiladas. I still don't understand how the supply of red chili never seemed to run out.

Some school nights, Little Mike and I would get to spend the night since the middle school we attended was within walking distance. Cholla and Sambo were early risers, so bedtime to them was the minute Mom and Dad left.

"*Vámanos*, Lawrence Welk is almost over," Cholla said.

"But it's only 6:30, Cholla," I said.

"We have to get up early," she answered. "Where's my *vaceta*?"

"Mike and I don't," I said. "You go to bed, and we'll be there after we watch Star Trek."

"Eeeeesh, I don't like those stupid shows. Get to bed now," she said, her Mexican accent thick. "We have to go to bed now!"

"Awww Cholla," I moaned. "Just a little longer."

"Right after your show, we have to go to bed. Don't make me get the *vaceta*."

Cholla brandished her vaceta like an Australian buckaroo with a stock whip. As I pushed her, she would leave a stripe or two on me in an effort to curb my rebellious spirit. Little Mike would get a couple from time to time for good measure, but it was more out of his unfortunate association with me than it was from his disobedience. There was more than one occasion when the *vaceta* would end up missing only to be found later in the back of the cabinet under the kitchen sink or maybe under the edge of a rug. Since there were worse things that could be brandished for doling out discipline, hiding the *vaceta* was a risk. Sometimes the enemy you know is better than the one coming behind it.

Sambo would watch it all in silence, then get up from his end of the couch just before seven o'clock, have a shot of whiskey, and silently go to bed. Cholla was left to deal with her hoodlum grandchild alone. We would end up in bed well ahead of when we wanted, but way later than Cholla wanted. It was one of those lose-lose situations.

There was a wall-mounted propane heater in the living room next to my favorite chair in their house. The front of the heater had a grill with horizontal slats that blew hot air on cold bodies; it was the next best thing to a wood fire.

One cold morning, I was getting dressed in front of this wonderful heater. I bent over to pick up my underwear, and my naked buttocks encountered the heated slats of the stove, leaving nine or ten perfectly spaced branded lines burned across both butt cheeks. Even though that area had better pain tolerance than other areas from experience with principal's paddles and other spankings, getting branded was quite painful. I went to school that day concerned about how I was going to sit through it, but that turned out not to be my biggest problem.

"OK boys, two more laps and we'll be done," Coach Tim said as we finished PE.

We trudged around the heated gym a couple more times and headed to the locker room.

"What the heck happened to you?" Eric asked me in the shower.

Of course, every other of the ten boys had a look.

"I burned myself on the heater this morning," I said.

My explanation didn't bring out the sympathy of my twelve-year-old peers; instead, there was hysterical laughing and pointing.

"Looks like you got grilled," one of them laughed. "Two all-beef patties."

"Now you're branded like a cow," from another.

I finished showering and went about my day. It wasn't long before even those who hadn't seen it were talking about it. Those lines stayed marked across my buttocks for several months, but the random mention of them from my associates lasted longer.

17

TEACHING AT AN ANGLE

My dad was a very smart man. His intellect has made his way straighter throughout his life, and it has been interesting to look back on some things that he did and realize what a clever teacher he was.

I hated shaving. I still don't love it, but as a rebellious youngster of twelve with more facial hair than I should have had, I really hated it.

"I had a friend in the Army who used to really lather up his face with warm water and soap before using shaving cream and a camel-hair brush," my dad told me one morning. "His name was Bob. He was from Michigan. Great guy. Knew a lot about shaving, being from Michigan and all. Anyway, you know what shaving can do to a face in cold weather, so he had this whole program. I'll show you if you want to try."

"OK," I said.

After all, if Bob did it that way and not my dad, well, it had to be better.

He went through a small ritual of washing his face, lathering with soap, and applying shaving cream with a camel-hair brush.

"Wow Dad, that really is better," I had to agree after trying it.

It didn't dawn on me that this was pretty much the exact same thing he had tried to teach me earlier, only in the first teaching, it was coming from my dad. Not from the imaginary, mystical, magical, Army buddy, Shaving-Bob who knew more about shaving than most because he was from Michigan.

I have found this technique to be very helpful throughout my career as an educator. For some reason, people seem to take instruction better with credit being given to someone else. There is also the added benefit that you can criticize performance by saying something like, "That looks really good, but when Bob did it, he would..."

Let me also add a word of caution. This technique can be overapplied. It is also 100% ineffective, perhaps even detrimental, if you try using it on your wife with your own mom as the comparison. Even using her mom as the comparison will not allow you to escape your mistake of bringing it up. You've been warned.

18

VETERINARY DREAMS

I had thought from a young age that being a veterinarian would be a great occupation. Of course, my folks agreed that this would be a good path to follow, so Dad made arrangements for me to spend some time with a local large animal vet the summer I turned twelve.

"Welcome," Dr. Ken said. "I'm going to have you with Eduardo today."

With that, he led me through the clinic and out to the stall barn where sick animals were kept. A Mexican man was raking a stall as we walked up.

"*Eduardo. Este chico es Cris, el hijo de uno de mis clients. Él te va a audar hoy,*" ('Edward. This boy is Chris, the son of one of my clients. He is going to help you today,') Dr. Ken said to Eduardo.

Eduardo was the quintessential Mexican and could have easily been the national symbol for what most Americans would have described to a sketch artist if they were describing a 35-year-old Mexican male. He was about 5' 6" tall, a little thick through the middle, with mahogany skin, black hair, eyes, and mustache, and was wearing an oversized straw hat. Put bandoliers full of bullets across his chest, and he would have fit perfectly in a 1960s western. Holding a guitar, he would look like the perfect mariachi band member, and his picture on a package could have sold tortillas or salsa.

"*Gracias, jefe,*" Eduardo answered.

"Chris, this is Eduardo. He doesn't speak any English, so I hope your Spanish isn't too bad," Dr. Ken said.

"I don't speak any Spanish," I said.

"I thought your grandparents mainly spoke Spanish."

"Yes sir, but not to me and Little Mike," I said. "Mainly just when they are talking to each other or to my mom."

"Well, you'll get along fine," Dr. Ken said, and he was gone.

"*Hola, Cris,*" Eduardo said. "*¿Comó está?*"

I shrugged and smiled. "Hi."

Eduardo smiled and nodded back. "*Ven aqui, ven aqui,*" he said.

I smiled back and just stood there. I didn't know that "*Ven aqui,*" meant, 'You come here.'

"*Ven aqui, ven aqui,*" Eduardo said, and he added gestures to indicate that he wanted me to come into the stall.

I walked into the stall. He handed me the rake and pointed at the ground where he had been raking the straw and poop into a pile. He left and I started raking. He came back with a wheelbarrow and an "apple fork" and started loading the pile into the wheelbarrow. An apple fork is like a pitchfork with more tines that are spaced closer together. It is designed to pick up horse "apples," a euphemism for horse manure.

Eduardo was a talker, and a fast one at that. He continued to keep up his end of the conversation even though all I could do was add the occasional nod, shrug, or "OK," into the mix. I followed him out to the pile where the wheelbarrow got dumped, and we continued through the eight stalls in the barn.

When we were done, Eduardo rolled a cigarette and offered it to me. I had to be as cowboy as he was, so I took it and we sat down on a pair of overturned buckets and had a smoke. As I said, New Mexico and America was a different place in those days and what I was doing would not have seemed exceptionally out of the ordinary. Mom and Dad would not have been happy about smoking, but they wouldn't have thought I was doing anything that incredibly bad either. Dad had seen Little Mike and I smoke Bull Durham with Pony and hadn't had much to say about it. Chewing tobacco was much the same. It was fine and dandy with adults as long as you dumped your spittoons and didn't leave them in the truck.

I handed Eduardo my Copenhagen. He put a little in his mouth and spit it out right away with a line of Spanish that had a few more words in it that I did recognize. He was not a fan of the snuff but didn't mind the Days' Work plug that I offered next.

I followed Eduardo around that day. We fed some sick cows and horses that were on the place, hauled some trash, fixed a gate in one of the barns, and pretty much did normal ranch stuff, even though it was a vet clinic.

"Hey Chris, your dad is here to get you," Dr. Ken said that afternoon, coming out back to find me.

I followed him into the clinic to go home with my dad.

"Well, how was it?" Dad asked as I climbed into the green Gladiator Jeep.

"It was just like working on the ranch, only with a guy that couldn't understand me, and I couldn't understand him."

"What do you mean?" Dad asked. "Did Dr. Ken use a bunch of big words or something?"

"I don't know," I said. "I only saw Dr. Ken about twice. I spent the day cleaning stalls, feeding things, and working with Eduardo."

Dad smiled. "Well, I guess you've got to work your way up in the vet world. You'll probably be castrating cats by next week."

My days continued much like that except that Eduardo and I got better at communicating. Not with words per se, but with gestures and grunts. He was a great guy and always had some sort of treat that his wife had made especially for me. He could do more with a rope than any man I had ever met, and he taught me a few rope tricks.

Dr. Ken knew that I was interested in maybe someday being a vet, so he included me when he could. I got to go on farm calls from time to time, and anytime something interesting happened in the clinic that I could observe, he brought me in.

"We're headed out to do an autopsy," Dr. Ken told me. "*Eduardo, no necesitas venir.*" ('Edward, you don't need to come.')

Eduardo smiled and nodded, and I followed Dr. Ken out to his truck. Dr. Ken's six-year-old son Timothy was in the truck, and he moved over to the middle as I opened the passenger side and hopped in. He had a bag of donuts from the Spic & Span Bakery, where they made donuts that I have never found the equal to.

"Wanna donut?" Timothy asked through a mouthful of sweet tastiness. He was holding the bag open with his chocolate-covered hands, having just freed one of them by stuffing half of a donut in his mouth.

I grabbed a donut and sat back to enjoy the change of routine as we bounced down the road.

"This guy has lost a few young calves in the last week or so," Dr. Ken said. "If it was only a couple I wouldn't be concerned, but his wife thought it was at least four when she called in. We're going to cut one open and see what we can find."

The place was close to the vet clinic, and we were there in a few minutes. The rickety fence that led away from the drive on both sides told the story of a place that was barely holding on financially. The natural scenery was beautiful with a mountain landscape and green grass coming away from a brook that flowed beside the two hundred yards of dirt driveway. When we got to the farmyard, there was a corral with a couple of old horses in it, the fence being made from some scrap guardrail, broken railroad ties, and a bedspring thrown in for good measure. The ancient trailer house had seen better days, and tires were placed strategically on the roof to hold the tin down. Geraldo came limping out to meet us as we drove in, his cane landing in time with his right foot.

"Thanks for coming," he said as Dr. Ken came to a stop.

"No problem, Geraldo. Glad we could make it. Where are the cows?"

"I pulled them to the top of the red pasture where we have always put the bodies. I'm sure worried Doc. All of them was young calves. Healthy and strong. Then they just died."

"I'm sorry, Geraldo. We'll have a look. Lead the way," Dr. Ken said.

"I'll get my horse," Geraldo said, walking toward the corral.

"You can ride in the truck," Dr. Ken said. "Boys, you sit on the tailgate."

He dropped the tailgate, and Timothy and I climbed on.

"I'd rather ride," Geraldo said. "I don't want to stay for what you gotta do if you don't mind."

"Of course," Dr. Ken said.

I could tell that Dr. Ken was getting impatient as Geraldo caught and saddled his old sorrel gelding, his limp making it take longer than it should have. He didn't want any help, so we stood there watching, Dr. Ken frequently checking his wristwatch. Timothy and I sat back down on the tailgate as Dr. Ken fired up the truck and followed Geraldo at a very slow pace to the red pasture where the dead cows were. Riding on the tailgate was always a treat and the first choice of New Mexico ranch kids everywhere.

"Right at the top of that draw," Geraldo said, pointing with his cane that he had kept with him as he rode. He had stopped his horse and was leaning in the stirrup to look at Dr. Ken in the truck.

"OK," Dr. Ken said. "We'll be back down in a little while."

Geraldo turned back toward the house, and we bounced on up to the carcasses. Dr. Ken was making up a little time, so Timothy and I were holding on for dear life. All of them were bloated and the stink was beyond anything I had ever smelled to that point. Dr. Ken looked disgusted, I felt sick, Timothy could have been at the park for all he cared.

"It's going to be a nasty one," Dr. Ken said, as much to himself as to Timothy and me.

He pulled on the long gloves that they use for pregnancy-checking cows and opened a scalpel as he headed to the least swollen of the four dead calves. Timothy and I followed, him with a bounce in his step and his bag of donuts, me with hesitancy while trying to appear bold, which apparently was not something that anyone can pull off.

Dr. Ken began by puncturing the distended gut. "Better to get the gas off when you know where it's going than to hit it later when you don't want to," he said, keeping up with his dialog that was directed to no one in particular.

The gas added a layer to the stink, but Dr. Ken soldiered on. Timothy had taken up residence by sitting on the head of one of the other dead calves, reaching into the bag for another donut. The donut that I had eaten was about to make its reappearance as well, except that it would be coming up the way it had gone down.

"Yep, I know that smell," Dr. Ken continued. "Let me cut this muscle and I'll know for sure."

He cut a chunk out of the top of the back leg. The muscle was speckled with tiny black spots.

"I better do one more to make sure," he said to himself as he moved past his son to the back of the cow that Timothy was sitting on.

Cutting that one, the muscle looked the same as the first one. "Smell that sweet odor?" he asked. "That's what blackleg smells like."

It didn't smell sweet to me. Not exactly. Maybe something sweet that had spoiled, and mixed in with a rotting cow? I haven't smelled that smell since, so I don't have anything to compare it to for you.

"Let's head back, boys," Dr. Ken said. "I don't think I'll need any samples."

He pulled one glove off, and then pulled the other glove inside out as he pulled it off that arm and made it into a bag to contain the stinky gloves. With that, we jumped in the truck and headed back toward the farmyard. Geraldo was sitting on a log in front of the barn that had been an old barn twenty years ago, his still-saddled horse tied to the rotted hitching rail.

"Did you vaccinate your cattle this year?" Dr. Ken asked.

"Well, I wasn't able to do it this year," Geraldo said.

"Had any of them calves ever had a vaccine?"

"No. I don't think so, now that you ask," Geraldo answered.

"What about any dirt work or new stuff going on around here?" Dr. Ken asked.

"My cousin has a dozer and did push up some dirt for a new pond this spring. But that's the only thing that's been done around here in a long time."

"Well, I think that is probably it then. Them calves died of blackleg. It's in the soil and they can ingest the spores which leads to their death. Mainly affects young cattle or very old cows that haven't been vaccinated. You should get the rest of your herd up and let me have a look in case you got more of this headed your way."

There was a small tear dripping down Geraldo's creased skin where his nose met his cheek. "I'll do it Doc. My wife will call when they're in. What do I owe you for right now?"

Dr. Ken looked away. "This one's on me," he said, almost too quietly for Geraldo to hear him.

No one said anything for a minute, the sad and awkward silence hung like the presence of a shade between us.

"OK then," Dr. Ken finally said. "I'll wait to hear from you. Have a better day. Sorry."

I felt like Geraldo looked. We all got back in the truck and headed back toward the clinic. Nothing was said as we drove, as best as I can remember. Timothy was out of donuts.

By the time the summer was over, I had seen quite a few interesting things and gotten a good taste for the life of a large animal vet. It was a fun and informative summer and I thought that being a vet wouldn't be a bad deal someday. Turns out I found a better path for me, but that was one I could have followed and enjoyed. It is one of those roads not traveled that we all have as we look back on our lives.

19

TECOLOTE CATTLE COMPANY

When I was thirteen, Dad secured me a job with the Tecolote Cattle Company (TCC). It was about ten miles of rough mountain and dirt road from our place, so I could drive to work without the likelihood of meeting one of New Mexico's finest. The J20 Jeep Gladiator (Green Machine) and I went back and forth every day like the old friends we were becoming. Driving on country roads as a thirteen-year-old was not at all uncommon in those days. Neither was chewing plug tobacco or snuff and getting a little Jim Beam in your coffee in the afternoon if there was snow on the ground and you'd been out working in it. A boy could even have a beer with the ranch hands on a hot day if he had been one of them. Different times than now.

The ranch was run by a cowboy's cowboy named Boyd. He was as tough a man as you have ever met and had lived the type of cowboy life that was common once upon a time but has become rarer as the decades pass. He could rope anything, train horses, ride a bucker, and was more comfortable on a horse than he was on his own feet. His winters were spent pulling calves and doctoring sick ones, feeding hay, and cutting ice for the cows to drink. Nothing about this lifestyle is as romantic as the movies might lead one to believe, and at thirteen, I was more interested in the things that my friends were enjoying. None of those things involved bad horses, frozen ponds, or freezing in the saddle while I checked cows.

Luckily for my thirteen-year-old self, the job was during school holidays and summertime, so I had the opportunity to hang out with my townie friends when school was in session. Their stories of sleeping in and playing all day made me jealous. The city life sure sounded like a good time.

As I remember it, I was paid $15 a day and all the coffee I could drink. My introduction was during the school winter break in December. That Christmas season of 1981, I would leave our ranch in the dark of morning. I would get to TCC by the time that the sun was making itself known, and I would cut ice in a couple of tanks and a pond before going to the ranch house where Boyd and his wife lived. He would be finishing breakfast, and I would get a nice big mug of coffee. I'd sit with Boyd as the other hands made it in, and we would get our orders for the day. Most days started with loading a flat-bed trailer with square bales and feeding the cows, then we would head into the timber and burn brush piles that had been made the previous summer by timber-thinning crews. In later years, I got to be in on the thinning during the summer as well. All in all, it was a very agreeable way to spend twelve hours a day as a youngster. It is better in hindsight than how I felt about it at the time, but it was instilling more of that work ethic that has served me well.

When summer released kids from the confines of school, I was once again working at the TCC. The first summer, I did a lot of fencing, and it was one of the toughest jobs I have ever had. There were a couple of massive college football players, Will and Mark, who were bosses on the fencing crew. They worked as bouncers on nights and weekends in nearby Las Vegas, New Mexico. This was a time in Las Vegas when being a bouncer all but guaranteed a fight, which suited those men just fine. Both were strong and loved to use their bodies as if they were all motors, cylinders, and hydraulics made by Caterpillar.

They used their livers hard every night, as well, but they had the stamina and fortitude to work through the hangovers.

There were new guys hired every Monday on the fencing crew. Sometimes there were five or six on a Monday, but by most Fridays, it was often just Will, Mark, and me left. Will and Mark were still there because they gloried in the pushing of their bodies. I was still there because at that age, I didn't have command of my own destiny and my dad wouldn't let me quit. Believe me, if he would have let me, I would have. Thank God my parents had the foresight to raise me that way instead of letting my comfort be their primary concern.

Thirty-plus years into teaching horseshoeing as a trade, I have seen the shift in the generations through the students who come to my trade school. There are a lot of youngsters who have been allowed to quit in their youth, and it has become habit-forming for them. I suppose the cycle continues as it must. Hard times lead to strong men, strong men lead to easy times, and easy times lead to weak men, which once again will lead to hard times. Since each of us live only the years God gives us to see it happening, we are surprised at how the generations change and history repeats itself. I think every old man has looked back, like I am doing now, and wonders what the future will hold. I suspect that you only need identify the part of the cycle you are in at the time to see what's coming. The weak men I am seeing now are sadly and unavoidably leading to some hard times. The main solace I get is knowing that it is all in God's hands.

There were days on the TCC where we would spend all day just hauling posts or wire into the mountains on foot from the closest place we could get with our beat-up Bronco and pickup-bed trailer. We would haul stuff for one day so we could build fence for two. It is difficult to say which of those days was harder.

Some of the workers recruited to work on the ranch provided some laughs. One city kid named Ty from Chicago had come to Las Vegas to play basketball at Highlands University. Ty was conned by Will to come and work on the fencing crew, and I don't know if Will did it for a laugh or if he really thought Ty would be a good addition. He was perhaps the tallest man I had met at that point in my life, so even though he was with us less than a week, he did get stuck lengthwise in my memory. On the very first morning, Ty got locked in the Bronco.

"Let's get going, Ty," Will said to the overly long fellow sitting on the passenger side of the Bronco. We had just arrived at the base of the mountain where we would be hauling stuff up all day. Ty was in the front seat because there was no chance of folding him up enough to sit in what was left of the backseat. Mark barely fit.

"I can't," Ty replied. "This door's locked or sumpin'."

"Try reaching over and see if you can open it from the outside," Will said. "The inside handle ain't never worked."

I had already jumped off the back of the trailer where I had been riding with a few other hands and was shouldering some steel fence posts, my leather gloves perched on my shoulder for protection instead of on my hands. Five posts to a bundle, two bundles were my designated load since I was so scrawny. We all stopped to watch the conversation Will and Ty were having.

"Wha's should I do, man?" Ty asked. His accent was hard to understand, but the scared and pleading look in his eyes wasn't.

"Well, if you can't open the door, just stand up and step over the door," Mark said. "Your legs are long enough to step over the whole dang truck if you wanted to."

The top had been cut off of the Bronco with cutting torches after it had been rolled a few years prior. It was like a hillbilly convertible that could not be converted back. There were no windows or top on it. It made the perfect little ranch wagon before side-by-sides became popular, except without a windshield, so there was always the potential of swallowing a bug if you pushed the speed that it still possessed. That old truck looked more like a busted-up white Army Willy's Jeep than it did a Bronco.

"Youse sure I can get out like dat?" Ty asked.

We all laughed and exchanged looks that didn't need words to interpret. I guess Ty was a great basketball player.

The fact that Ty couldn't figure out how to get out of the Bronco spoke volumes about the quality of the guys who would start on

Monday and be gone by Friday. It got to be a game to see whom Will and Mark would recruit to work that fence crew and how long they could last. At the end of the summer, Will, Mark, and I were the only ones that had made it for the whole season. Again, thank you, Mom and Dad.

20

BRANDING TRADITIONS

The testicles sizzled and popped on the hot rock. "Ready to eat," Will said to Mark and me as he tossed us each a half-cooked Rocky Mountain oyster, fresh from the calf.

The testicles were encased in a tough membrane that was removed when cooked properly. When Mom made them, they were dipped in eggs and then rolled in salted flour and fried. Delicious!

At a branding, "cooked" meant that they were set on a rock until the tough membrane split from the heat. Only the area where the split occurred would be cooked enough to be considered edible by anyone other than a sushi chef, but Will was neither a discerning nor talented cook.

Eating Rocky Mountain oysters was a branding tradition. The flanking crew ate the first few that had been set free when my dad turned the bull calf into a steer. There was a fire going to heat the branding irons in the hot and dusty New Mexico pen where we were branding. Some brandings had homemade branding-iron ovens that burned propane, which my dad did eventually get, and it made for a more thoroughly cooked testicle. At this time, the flanking crew was using a hot rock for their grill because Dad was using wood for his branding-iron fire.

Spring was branding season in New Mexico, and Little Mike and I did our share of "neighboring," which means that you helped your fellow ranchers with their brandings. They returned the favor when it was time to brand at our place. The morning started early with gathering the cattle and penning them, followed by roping and branding the calves. Boyd brought Will and Mark over from the Tecolote Cattle Company, and RD had brought some of his unnamed transients.

I didn't even get a chance to rope when it was our home branding. Little Mike was the magician with a lasso, so it fell to him and Pony. I was allowed to rope at a few of the neighbors' brandings until I brought a calf to the crew with only one foot caught, or, heaven forbid, a neck catch. Doing that put you on the flank crew in short order, but I fit in with the flanking crew better than I did with the ropers. Kind of like "roughies" at a rodeo compared to the team ropers. Little Mike was like Will Rogers with a rope, and he got to stay mounted longer than most.

"Mine's still raw," I said to Will.

"Just eat it," Mark said. "Here comes another."

I bit the cooked bit where the testicle had split and tossed the remainder back on the hot rock. Still chewing on the part that I had bit off, I grabbed the rope on one side and Mark grabbed the tail on the other side as my brother dragged the next calf between us. As the calf hit the ground, I grabbed the top hind leg and sat down, pulling the leg toward me with my foot on the bottom hind leg. Mark put his knee on the calf's neck and folded the top front leg around to help hold the calf down.

In this position, it was attacked like ants on a grasshopper. My dad came by to see if it was a bull and needed to be castrated, another couple of guys brought the hot branding irons to burn an S, a P, and a bar into the calf's side to make the San Pablo Ranch brand. Dad made a swallow-fork mark in the ear using the same knife he castrated with, and other hands were administering shots.

Little Mike had ridden Corky back into the herd and was dragging another calf between Will and Jake, the other flanking crew. He had beat Pony to the next one, but Pony was riding a colt and really wasn't a match against my brother when it came to roping.

Mark and I let our calf up, and I went over to the hot rock to see if my testicle was done cooking. It had been grilled to perfection, and I smiled as I finished my morning snack. It was not nearly as good as the ones my mom would make for me later, but it was enough to help get my hungry body through the morning. Life was good, but I didn't really know it at the time.

21

HIGH SCHOOL FRESHMAN AND NORM

By my freshman year of high school, I was starting to really identify with the "stompers." We were the ranch-raised kids who wore boots and cowboy hats, chewed plug tobacco or dipped snuff, and were more into rodeo than football. With this identity came taking Ag class (Vocational Agriculture) and being a part of the Future Farmers of America, or FFA. Even though I had played football and been on the track team throughout middle school and during that freshman year, I was not a gifted athlete and did not really fit in as well with the jocks, especially after spending my first full summer working for the Tecolote Cattle Company. I was as fit as I had ever been and brown as a beaver, but I was gaining more interest in rodeo than I was in team sports.

One of my best friends, Tim, was also in FFA with me, and we got into a world of trouble together. The shop class was run by a short and supremely muscled little man named Mr. Scott, who had graduated from New Mexico Military Institute (NMMI) – a legendary place where I was destined to be going.

There is a tradition at NMMI when cadets get initiated into the Corps of Cadets and get their epaulets on their uniforms. It involves something called a "zoom broom" and a lot of beatings with said zoom broom. The zoom broom was just a normal broom that had the handle cut down to about three feet and the bristles wrapped in grey duct tape to give them a board-like rigidity. On initiation night, Mr. Scott was overly abused for his height, or lack thereof, and took it upon himself to become a master of the zoom broom. This skill easily translated to paddling juvenile delinquents that were in his Ag class.

"This is the 1982-83 shop paddle, named Norm, and you will have the honor of signing Norm every time you get swatted," Mr. Scott said.

It was the first day of shop class, and Mr. Scott was holding up the finely crafted piece of wood that he had carved into a paddle. It was two feet of finely sanded and varnished pine, with a tapered handle, said handle being wrapped in cloth tape for extra grip. "Any problems in class will be handled in class," he said.

"What sort of problems?" Tim asked.

"Raise your hand if you have a question," Mr. Scott said, pointing Norm at Tim.

Tim raised his hand. Mr. Scott said, "Do you have a question?"

Tim asked, "What sort of problems?"

"In other classes, the teacher will often send a student to the principal for fighting, talking back, disrespect, cursing, vandalism, or other offenses. Norm and I will not be needing the principal."

"Yeah, right," Tim laughed. "You think you can swat us just like that?"

"In my office, now," Mr. Scott said.

Tim looked surprised but didn't move from his seat. He looked over at me. I shrugged since I didn't know what else to do.

"Now!" Mr. Scott repeated. There was no room for disagreement in his tone.

Tim got up slow and nervous-like, then walked toward Mr. Scott's office, which was situated between the classroom and the actual shop where all the welders, grinders, and other tools were located.

The door closed as Mr. Scott followed Tim into the office with Norm over his shoulder, carrying it like a baseball player headed from the dugout to home plate. We heard a loud whack and Tim came out of the office, red-faced and trying hard to keep from letting tears escape.

"Are there any more questions about Norm?" Mr. Scott asked.

Not a hand went up, and no one acted like the teenagers that we were.

"Very well. By the way, young man, you get to be the first to sign Norm," Mr. Scott said, bringing the paddle and a marker to the table that Tim and I shared.

Tim physically shrank back from Mr. Scott and Norm, and his signature was decidedly shaky.

"Man, did that hurt bad?" I asked Tim as class ended and we headed toward the main school building where the other classes were.

"That little #$*&er can really swing a paddle," Tim said. "I'm going to get him for that."

"What are you going to do?" I asked.

"I don't know yet, but that was uncalled for."

"I think he wanted to make a point and set an example. You just happened to make it easy on him," I said.

"I didn't even do anything wrong," Tim said.

"You have to admit that everyone was sure on their best behavior after you got whacked."

"Maybe, but he's going to &*(%ing pay for that," Tim said. "Totally uncalled for."

Shop class became our favorite class, despite the teacher. He had his good moments and later told stories about NMMI that explained his technique with the paddle. Tim and I enjoyed the class mostly because of the things we were learning and some of the mischief we could get into.

We would do stuff like putting grease in a welding glove, on a hammer handle, or a thin layer of it on the glass of the welding helmets, or loosening the welding leads from the red Lincoln "cracker box" welder just enough that a student couldn't figure out why the welder wasn't working. If anyone was caught touching a welding table and grounding themselves, you could touch them with a welding rod and give them a heck of a shock. I endured the pranks as often as I gave them out. Getting caught doing stuff like that would mean you ended up with your signature on Norm, and I did get to practice signing my name on many occasions, but not as often as Tim, since he was awarded the paddle at the end of the year.

22

TIM'S PAYBACK

It snowed one day, and Mr. Scott parked his beige-colored Chevy pickup truck in the shop. "I've got an idea," Tim whispered to me, coming up to the welding table where I was working on laying down a bead.

"What's that?" I asked.

"I'm going to weld his lug nut to his wheels," Tim said. "We'll get Dan and Bruce in on it and look for an opportunity."

I smiled to myself, thinking of all the times I had passed a marker over Norm. It would require some teamwork, moving the welder into position, removing the hubcap, and some simple, quick tack welds, all while not getting caught. I regret to write that I was very much in favor of this prank.

The class went on and Tim had set the plan in motion so that those needing involved knew their play. I was going to move the cracker-box welder and ground the truck. Dan would remove the hubcap, and Tim would do the welding. Bruce would stand as the lookout in the office between the shop and the classroom, hopefully giving us enough time to be in the right place when Mr. Scott got back into the shop. There was no chance that first day, and for the next several times that the truck was parked in the shop, the opportunity did not arise. Tim was even thinking of sneaking in at lunch or maybe even causing a disturbance in the school to get Mr. Scott out of the shop. By the time we actually got the prank done, he had devised about a dozen different plans to make it happen.

As spring came to New Mexico, we finally had an opportunity.

"Sam, I have to run to the main office for a few minutes. Keep an eye on things for me," Mr. Scott said to the teacher's pet.

"Yes sir," Sam said, lifting his welding helmet.

Tim, Dan, Bruce, and I all exchanged looks. This was what we had been waiting for.

"You had better just keep welding and keep your hood down," Bruce warned Sam. "Nothing's happening here that you need to know about."

We had been worried about Sam from the start, but Bruce was big enough and scary enough that we were pretty sure Sam would behave. Turns out we were right. Sam dropped his helmet and went back to welding.

Bruce went into the office and stood at the door that led to the classroom. I drug the welder over and clamped the ground to the bumper as Dan popped the hubcap off. Tim was tacking the lug nuts to the wheel rim, zip, zip, zip.

"Let's do another one," he said through his laughter.

"We ain't got time," Dan said.

"Hurry," I said. "Just one more."

Dan popped off the front hubcap from the driver's side and Tim moved up to continue applying his welding skill. Zip, zip, zip.

"Let's do the other side," Tim said. His confidence was riding high.

"Hurry guys," Bruce hollered from the classroom. "I hear something."

That was enough to get us back to our original positions and not move on to the other wheels. It seemed like hours until Mr. Scott returned. Bruce had been hearing ghosts. We would have had time to weld all the wheels completely, but I think Tim had done enough.

I never heard about what happened when it was time to change a tire. Did Mr. Scott have a flat he had to deal with, did the tire shop have an unwanted surprise, or did Mr. Scott discover it and take care of it before it was a problem? I guess I will never know for certain, but it made the whacks I received from Norm feel a little more deserved for the rest of the school year.

23

LA BOHEMIA WINE

I stand to be corrected on the name, but La Bohemia is the name I remember for the cheapest and sweetest wine that could be had in New Mexico in 1983. I had left school with Mr. Scott and Alejandro around ten that April morning. They were both seniors and we were going to go fishing at Story Lake to relax from the rigors of school.

"I got some LB," Mr. Scott said. He produced a brown paper sack from the floorboards of his F250 truck that he was driving.

As a freshman, I was sitting "cherry," or in the middle of the bench seat because there is an unspoken rule about that if you're the junior man in the truck. That seat was often a hassle in those days when every truck of that vintage was a manual transmission and the person sitting in the middle seemed to never be out of the way when the driver needed to shift.

"I'll get us started," Alejandro offered. He opened the bottle and took a swig. "Good stuff right there."

"Yes sir," I said, taking my gulp as the bottle passed me on the way to Mr. Scott. He took his swig and passed it back to me. Every time the bottle passed between them, I would take my share. Like the government collecting tax. The problem, besides the fact that I was drinking in the first place, was that my position in the truck meant that I was getting half again as much wine as Mr. Scott and Alejandro.

We fished, although I don't remember much of it. I do remember that Alejandro wanted to get back to school after lunch because he had a class where he could see a girl named Niki. He thought that seeing her might lead to getting to touch her. From what I know, it never did.

I found myself in Civics class at some point after lunch. My head was swimming, and the room was spinning. I got up and left the classroom in a dash, headed to the bathroom. Before I could make it, I lost some of the wine in the hallway, coming back up the way it had gone down. I continued to the bathroom, where I sat down in a stall and tried to get the nausea to pass.

The Civics teacher was Mr. Sanderson, and he knew without a doubt what had been going on. I had often left fish on the doorstep of his house when I had been fishing instead of attending his class. Since my grades were good, he probably figured that there was no harm done. It was like an unspoken agreement that we had.

"Did you do this?" I heard a man ask in the hallway.

"What, no sir," a kid answered back.

The man's voice asked again, "Have you been drinking?"

"No, no," the kid answered.

"You look like you've been drinking. Tell me the truth or you'll have to go to the office and see the principal."

"Well, maybe a little," the kid admitted.

"You need to clean up your mess right now," the man said.

"But I didn't throw up," the kid said again.

"It had to be you," the man said. "Who else could it be?"

"It wasn't," the kid said.

"Do you want to go to the office for drinking?" the man asked. "If not, I suggest you clean this up right now."

"But I didn't do it."

"Go to the office."

"I'll clean it up," the kid said.

He burst into the bathroom grumbling about having to clean up puke in the hallway that he didn't do. "Stupid teachers making me clean up someone else's mess. It's not fair. What the heck," he said as he jerked towels out of the dispenser.

I just lifted my feet above the sightline of the stall door and stayed quiet. Mr. Sanderson winked at me when I got back to class. He had followed me out when I was running to the bathroom. It had been his voice that I had heard getting another, less drunk, student to clean up the hallway of Robertson High School.

After church on Sunday, we went to the Dixie store on the main street in Las Vegas. This was like a modern Dollar General. We were headed down an aisle when I saw the familiar form of Mr. Sanderson at the end of the aisle.

"Isn't that Mr. Sanderson?" Mom asked me.

I tried to pretend like I didn't know who was at the end of the row, but he turned and headed our way. I just knew that I was about to be in a world of trouble.

"Hi, Mr. and Mrs. Gregory," he said.

"Good morning," my dad said, shaking his hand.

They made small talk while I stood there trying to look invisible.

Within a minute or two, we had said goodbye and were headed about our business. I looked back over my shoulder and Mr. Sanderson winked at me again. He continued to get every fish that I could catch for the rest of the year.

24

SUMMER OF 1983

"You have a break from the fencing today, boys," Boyd said as we stood on his porch to get our orders for the day. "The big bosses are coming in from the city, so the fence crew is mowing the lawn and cleaning up around the big house."

That sounded like a vacation day to me. Will wasn't all that happy about it. He would much rather attack a fence line with an axe or haul rolls of barbed wire and T-posts up the mountain. Yardwork wasn't ranch work, and that man was all about doing stuff that cowboys did. For the record, cowboys aren't any happier about fencing than Will was about yard work.

Like most ranches, the tack room was full of the finest saddles and bridles, kept in good repair. The leather was oiled, saddle blankets were on racks that would allow the sweat to dry, and all the bridles and halters were on their designated hooks. The tools needed for yard work, well, that was another story. These were shoved between a stall panel and the back wall of a barn, covered in cobwebs that were as much dust and bird poop as they were spider silk. It was a great place to find a rattlesnake as well, even though there wasn't one there that day.

In New Mexico, the demand for a large and usable assortment of implements for yard work is pretty small. We had an ancient push mower at our ranch that didn't have a motor. The faded rose-colored frame was probably once bright red, but the sun had taken care of that. At least rust wasn't a problem. They are called reel mowers, and the blades spin from the geared action of the wheels being turned when pushed across the ground. Not that easy to use, nor do they do that good of a job, but when you only have to mow twenty square feet of grass in the yard just twice a year, it is good enough.

TCC did have a push mower with a motor, so that was snazzy. There were a lot of pinecones and needles in the yard, so it was more brown than green. There were areas where the grass was fighting for dominance, and after we had been over the whole thing with the mower, I was assigned a rake.

The handle in my rake was a piece of thin metal tubing that had been shoved into the head of the rake and secured with a screw and some baling wire. The top of it had been shortened by bending it in half and working it back and forth at the bend until the metal gave up. This left the end flat with sharp corners on both sides, much like the end of a two-inch wide chisel. Using it was fine if you avoided the top of the handle and didn't get too horsey with the rake end.

"I can't believe we're doing lawn care," Will grumbled to Mark after lunch. His mood was still dampened by weed eating and mowing.

"I'm not too disappointed," Mark said. "I could use an easy day after that fight last night. I think one of them drunk SOBs must have clipped me on the side of my knee."

"Yeah, that was a great one last night," Will smiled.

Mark had cheered him up a little by bringing up the happy memory of the bouncer bar fight they had participated in the night before. They had a good story about once a week, and sometimes they had some interesting wounds to enhance the telling.

"I did see that now that you say it," Will said. "I think that bad man hit you with a pillow, you big baby."

Mark smiled and punched Will in the shoulder hard enough to knock down a normal man, but it was only enough to make Will smile bigger. Being amongst them, I felt like you sometimes feel when you are feeding cows or horses in a feed trough. Just avoiding getting trampled and ended by the giants around you becomes paramount.

Will grabbed Mark and they started to wrestle. Mark was a college wrestler that had almost qualified for the Olympics. Will was football-strong and football player-sized, but Mark had the technical advantage and soon had Will pinned. They were both laughing and enjoying themselves, and the mood had brightened. They didn't wrestle often, especially since the fencing demanded so much sweat equity. Doing yard work had made their child-like energy spill over.

Lunch was done and I was raking when Will grabbed me from behind. Mark hadn't completely emptied Will's energy, and I would be an easy conquest after getting pinned. I jumped and as I came down, the sharp end of the rake handle went into my right eye. There was blood and immediate swelling.

"Oh man," Will said. "I guess you'd better head to Boyd's house for the rest of the day."

It was only about a quarter mile, so I walked down to Boyd's place. His wife was just getting home from her job in town.

"You can just rest on the couch until Boyd gets back," Joanie said, giving me a plastic bag of ice. "He's at the white house mesa checking on the cows in that herd."

"Thanks," I managed.

I fell asleep on the couch until Boyd made it back close to quitting time. By then the eye was completely swollen shut and hurting quite badly.

"You think you can drive home?" Boyd asked.

"I don't think so," I said. "I might call my dad so they can come to get me."

"Sounds like a good idea," Boyd said.

I called my folks and explained what had happened. Since I was only fourteen, driving the dirt roads to the ranch was acceptable. Driving to town or anywhere that was busy would have been out of the question. Mom and Dad showed up and Dad pried my bloody and swollen eyelids apart.

"Your eyeball is cut," he said. "We better get you to town."

That sounded like bad news to me. We went to the hospital in Las Vegas, but they sent me on to Santa Fe. The main eye surgeon said that I had broken my orbit under the eye. The recommended surgery was to place a Teflon plate under the eye where it was broken so that the muscles would continue to work normally, and I would not end up with a lazy eye. He pointed out the fractured bones that were easily seen in the x-ray.

The doctor was also concerned that I may end up losing the eye, and he gave me a few exercises to work on such as trying to touch certain places on a colored poster that you set flat in front of your one good eye. Depth perception is something that I very much had taken for granted.

Mom had set up a huge prayer chain amongst her army of prayer warriors. It was a God thing because when they opened me up, I didn't have a broken orbit. My eye healed with no problem, and I just ended up with a week off and a story to tell. I sure looked like I had survived one heck of a fight when I got back to work at the TCC. I had a big black eye for weeks and the white of that eye became solid red from the broken blood vessels.

My eyesight in my right eye was better than in my left eye after the injury. While I would not recommend that injury, when God healed me, he gave me a little extra in the bargain.

I officially left home to attend New Mexico Military Institute in August of 1983, having been fifteen years old for just over a month. Even though I chose it voluntarily and with a full desire to go, there is no doubt that my behavior could have warranted my being sent there. NMMI constructed things in my mind and personality that would not have been built had I remained in Las Vegas.

While I still enjoyed holidays and summers at home until after I was eighteen, the truth is that things are never the same after a youngster leaves home for an extended period. I was still the son, and they were still the parents, but the roles change even when the actors do not. I should say that the actors are the same people, but experiences have changed them, as they do all men. So, I guess the actors do inevitably change. Since my folks were farsighted enough to realize this, it was not hard to live under their roof again after having been out in the world. They had enough respect for the man I was becoming to let me make the mistakes needed to teach me the lessons that were also needed. There was also a vast array of mistakes that they were not privy to.

Once I became a parent, thanking God it did not happen by accident, I was able to draw on the example set by my wonderful models. Like my folks, my intention was to raise independent and free-thinking adults. I, too, was similarly surprised when my offspring were ready to apply that free-thinking to their lives and leave the comfortable nest Kelly and I had made. Such is life, and I have come to see God's intention in it.

PART II

THE INSTITUTE

25

FIRST TASTE OF NEW MEXICO MILITARY INSTITUTE

"Hi, Mr. Gregory," Sgt. Farr said as my family and I stood in the Sally Port of Hagerman Barracks at NMMI in August of 1983.

He was deceptively kind and polite, no doubt having been chosen as a greeter by the cadre for his ability to mask the horrors that awaited.

"You are in Delta Troop, cadet number 830087," Sgt. Farr continued with a nice, but lying, smile. "It is just this way if you will follow me."

With my mom, dad, and brother in tow, we followed Sgt. Farr to the Delta Troop area of Hagerman Barracks, otherwise known as The Box, which was to be my home for the next nine months. The Box was three stories high, arranged as a great rectangle made of yellow bricks, with the dark green doors of the rooms opening to the inside of The Box. The Box provided housing for 600 men divided into ten companies (called by the cavalry term, troops, at NMMI) of about sixty to seventy men each as well as space for guard boxes and offices. In total, the barracks was 102,000 square feet with 306 rooms. In front of all the rooms on each level was a long porch, known as the stoop. High school-aged new cadets lived on the third stoop, most of the older high school-aged cadets lived on the second stoop, and the bottom stoop was primarily for college-aged new cadets. Each stoop had a concrete walkway that led to the center of The Box. Each of these walkways was large enough to form up a full troop of cadets. These concrete walks were known as troop walks.

How I had come to be here was an accident of fate that was borne of the great knowledge and perception that all teenagers have as a rite of passage at that age.

I wanted to rodeo, and NMMI had a great high school rodeo team with a great coach. All that I knew for certain was that I was going to get to participate in that sport as a part of school, and the rest of what went with going to a military school escaped my attention. I was not even sure they wore uniforms when I started my pursuit, and I was equally ignorant of the fact that I would be subject to the rules and laws of that academy since I was only there to rodeo.

As Sgt. Farr led my family and me closer to Delta Troop, I began to get the feeling that it might not be just an easy life of school and rodeo. On the bottom stoop and lawn in front of every troop, bags were strewn around, kids with shaved heads in civilian clothes were doing push-ups, and there were a lot of uniformed cadets that seemed mad about something.

"Front and center," Staff Sgt. Glaser said as I walked up to the stoop.

The joyful, jovial, and deceptive Sgt. Farr had disappeared to snag the next hapless victim.

"I don't know what that is," I said.

"I DON'T KNOW WHAT THAT IS, STAFF SERGEANT!" Staff Sgt. Glaser hollered, inches from my face. His thick German face was red and beaded with sweat, his blue eyes filled with rage at my stupidity. "STAND HERE. LOOK THERE!" he pointed with one hand to the edge of the concrete for my feet, and his other hand at the bricks of the barracks for my eyes. "CHEST OUT, EYES FRONT, CHIN IN, HEELS TOGETHER, AND STAY THERE."

I moved into position and heard sobbing from my mother behind me. I turned to look and could see the trace of a smile on my dad's face and a look of sadness from my brother. My dad had explained to me that I didn't want any part of going to military school. When I pointed out his obvious ignorance and the availability of rodeo, he finally relented. He had served in the US Army as a Field Artillery officer during the Vietnam Era, so he knew more about the military than I did, even though I was a teenager and knew it all. Once my dad had agreed to let me go, he made it clear that I could not quit. From my vast knowledge and expertise, I thought that was a trifling point that did not hold any relevance to me.

As I turned to look at my mom, Staff Sgt. Glaser hollered, "EYES FRONT, MAGGOT."

When I did not comply immediately, he once again got in my face, repeated the command, and made certain that I understood his word was law, adding some spit spray to his orders.

Turning to my folks, Staff Sgt. Glaser transformed as quickly as your mom does when she is yelling at you one minute, then the phone rings and she answers it like you were all sitting by the pool sipping iced tea. With a pleasant tone, he said, "Mr. and Mrs. Gregory, thank you for bringing your son. I will be his mother and his father for the next nine months, and you will be able to talk to and see him at Thanksgiving. You may go." He smiled sweetly as my folks turned to leave.

With that, they were gone. I was only six weeks past my fifteenth birthday, and my world had just undergone an immense change that was to affect the rest of my life.

26

RAT RACE

As soon as my folks were gone, the training of Cadet Recruit-At-Training (RAT) Gregory, C.L., commenced. My bags were thrown on the ground and some of the Delta Troop cadres began to go through them looking for contraband. Others joined Staff Sgt. Glaser in getting in my face and yelling at me. When my eyes moved even a quarter of an inch, I was dropped for push-ups. My shoulders weren't back far enough, so they physically pulled them. There were so many orders coming at me that I couldn't take half of them in. All around was chaos as many other RATs were being subjected to the same. As soon as there were four or five of us with our disheveled bags thrown in our respective rooms, we were formed up and marched single file to the barbershop.

It is NMMI's strategy to shave a new cadet's head as soon as he arrives. More than 100 of these newly shaved boys will quit and flee in the first week which is known as Hell Week, but they won't get to return to their civilian life without the telltale sign of a hasty head shaving. And I do mean hasty. For many, this embarrassment alone is enough to tip the scale in favor of them staying. For others, their quit runs deep and strong, and they can invent a reason to explain their hairstyle or wear a hat.

We hit the barbershop to stand in a line of over twenty other RATs from other troops that were all waiting for one of three available black-leather barber chairs. It was a classic-looking place, with light green walls, a black and white checkered tile floor, and a full mirror behind the shelf that held the implements of the barber trade. The three barbers were a kindly-looking older American woman, a short, thick Mexican man, and a nondescript Caucasian man. They had sweated through their barber's aprons in the excitement of having RAT-race haircuts on the boys in their respective chairs. Though the line looked long, it did not take long. Soon I was in the chair of the kindly-looking older woman. She smiled at me with a caring look that hid the truth. When the Mexican barber said go, she hit my forehead with the clippers and burned a path over the top of my head on the way to the back of my neck. Repeating the move as she raced over my nervous scalp, and the hair fell away from me like a waterfall. When she got to the side of my head, she grabbed my ear and jerked my head around so that the hair could meet the clippers with greater speed. I have never since been so competently and rapidly shorn.

I was unaware at the time, but she was a celebrated speed champion RAT-shearer. Her great swiftness came at the cost of a few RAT heads and ears, but that didn't bother anyone but the RATs. In mere seconds, my normal haircut was converted to that of a convict or state hospital inmate, and I was being ordered from the chair so that the next race could commence. Of course, she had won her heat in the round in which I had been an unwilling participant, and the Mexican barber reluctantly put another dash on the chalkboard under her name.

. . .

When I got back to my room later, I was relieved to see that my eyebrows had been spared, since the occasional loss of those in a RAT head-shaving race leads to an even worse look than just the top gone, as was evidenced by some amongst me.

With our new appearance, the five of us were lined up outside of the barbershop and marched back to the barracks. Being in step and avoiding looking around at the strange and new surroundings was not yet a part of our nature, so Cpl. Thompson kept up a barrage of insults and curses as he expanded on the "grodiness" and ineptitude of the incoming class, and the doubtful virtues of our mothers, which he claimed firsthand knowledge of. Cadets used variations of the term "grody" often; it means disgusting or foul, and was used in many ways, from calling a person a grode to the room being grody, with grodiness being the nature of a RAT.

There were all kinds of "RAT races." RATs could be raced against each other in virtually every way imaginable. Eating, running, packing a rucksack, going upstairs, going downstairs, moving rocks, police call, showering, etc. If it is a human activity, you can bet that at some point a couple of old cadets decided to pit one RAT against another.

27

PINK SLIP

Within hours, we had all done many push-ups, had run from here to there, been schooled on how to stand, how to walk, and how to talk (which was really not at all), and the cadre was beginning to affect how we thought. That is the essence of military discipline. Once the thought process of the young American civilian had begun to change, the process of making soldiers had started.

It was a frightening place. Having young men in charge of young men can be as rough a situation as there is. The old cadets showed no mercy as they made their charges bend to their will.

Soon, I was put into a group that was known as a squad. My squad consisted of nine people in the 2nd Platoon of Delta Troop. Four of us were high-school RATs, two were college RATs, and there were two team leaders and one squad leader that were in charge. After being assigned to our squads, we were marched to the PX (Postal Exchange) to be issued our uniforms. With the gear came a pink slip that detailed what we had received and what was out of stock. Since our parents had already paid for all of it, not having something only meant that we would get it later when we produced the pink slip for inspection at the PX. It was impressed upon us that we should keep a close hold of that slip.

THE GEAR WAS THRUST at us in a fashion that anyone who has been issued military uniforms can appreciate. You start with an olive-drab sea bag that is too small to hold half of what you are about to get, and things are piled into it and thrown at/on you in an assembly line manner. Each item was marked off the pink slip as it was shoved in amongst the pile of black combat boots, socks, shirts, brass, PT shorts, etc., and only those with longer-than-average arms could hold or see over the top of it.

WHEN I GOT BACK to my room on the third stoop of Delta Troop, I started to put all of my gear in my locker. As I went through the pile of stuff, the pink slip was not to be found. My dress green jacket and a few of the Class B shirts were out of stock at the PX, so they represented a fair sum of money in clothing that was due to me. I looked through everything that I had been issued, but it was not there.

ALTHOUGH I WAS nervous about it, I looked out of the window to the stoop outside. We had been ordered to stay in our room until called for, and we had not been called for yet. I saw a pink slip on the stoop only feet from my door. I don't have proof that it was for certain mine, but I have always believed it was. I stared at it for what seemed forever until an old cadet walked by and picked it up, gave it a look, and laughed as he threw it in the trash. Seeing my face in the window, he screamed at me for looking out the window.

. . .

So far, I had been at NMMI for about 3 hours, and I had yet to see the rodeo grounds. While you may be laughing at my ignorance here, I felt that there had been a bad mistake and somehow, I must inform the cadre that I was here for rodeo and not all the rest of the nonsense.

My roommate turned out to be a despicable fellow, and that was not good news for me. If you happen to be unlucky in the roommate draw, then unnecessary attention is thrown your way purely out of association. This is a fact my younger brother had suffered in civilian life as experienced teachers came to paint him with a brush that I had dipped in their paint. On that first day, I was unaware of the nature of my new companion, but I was certainly aware by the morning of the second. The cadre had figured it out ahead of me, and I found throughout our few months together just how miserable it could be to have a grode for a roommate.

28

FIRST DAY OF HELL WEEK

Within a short time, we were told to get into our PT uniforms, which consisted of black shorts, a red t-shirt, an NMMI cap, and tennis shoes. Once attired, the troop was formed up on the Delta Troop walk in front of our portion of the barracks and we were marched to chow. I was glad for this, as I had worked up a considerable appetite. That was a condition that plagued my entire youth, except for those few short and glorious minutes following a good meal.

As we marched, I caught the attention of a Sgt. Swan. When he saw me glance at him, he came over and pulled me from the ranks. The rest of the company marched on to the mess hall as I stood on the side of the road to talk to Sgt. Swan. I was happy that he had seen me since he looked like a nice guy, and I was sure that he could fix the mix-up with the whole rodeo thing.

"What are you looking at, maggot?" Sgt. Swan asked. "Are you a faggot that is in love with me?"

"Nothing sir, no sir, sorry sir, of course not," I answered. His nice-guy façade had disappeared with alarming rapidity.

"I am not a sir, you stupid #&%t," he yelled, his revolting breath assaulting my nostrils from an inch away. "I work for a living. Call me Sgt. Swan, the Stalker of Your Nightmares."

"Yes sir," I said again.

"Are you really that stupid?" he asked. "Get on down and pump this post to hell if you can't figure out the difference between a sergeant and an officer."

"Wait," I pleaded. "I am here to rodeo, not for the military stuff. Where can I go to get to the rodeo grounds?"

He laughed as he called over another of the old cadet cadre that had stayed behind to watch the fun that he knew was coming.

"This RAT isn't here for the military stuff, Charlie," Sgt. Swan said to a Sgt. Johnson.

"Is that right?" Sgt. Johnson asked me.

"Yes sir," I answered.

"I am not a sir! You stupid piece of crap. I work for a living. Don't you know what a sergeant is yet, maggot?" Sgt. Johnson said, uniting the aroma of his breath with Sgt. Swan's, again inches from my face. "Get on down, RAT."

I was weak from the push-ups I had already done and was on shaky arms after only a few reps. I was barely holding the front-leaning rest position, and the cadre was yelling at me, cursing my mother, despising the date of my birth, threatening my future children, and all in all scaring the heck out of me.

The rest of Delta Troop was long gone by this time, and the moments I spent with the old cadets seemed like an hour. When they finally let me up, they made me sprint back to the troop and get back into the formation. On the way, I saw a couple of my fellow RATs that had been singled out for whatever reasons the cadre could think of. Mostly these were the guys who happened to look out of the corner of their eyes as they were being marched to mess. The cadre was lying in wait to catch those who had the nerve and audacity to move their eyes.

We went into the mess hall, commonly referred to as the "Den of Culinary Atrocities." We were about to be introduced to how we were to eat two meals a day for six days a week for the next nine months. It was not the sort of way that you would eat if the object of eating was acquiring calories and nutrition, but it was the way a RAT had to eat. Mess-hall meals made every cadet realize that the food was of such a flavor that consuming large quantities was not suggested, either.

The mess hall was filled with long wooden tables with four chairs per side and a chair at each end. On one end sat the squad leader (mine happened to be Sgt. Swan, what luck), and on the other end was the A-Team leader. The other B-Team leader would end up on the right side of the A-Team leader when he got there, but at this time there was not a team leader for the 2nd Squad, 2nd Platoon, Delta Troop. For the sake of the cadre, that spot was being filled by Sgt. Johnson for Hell Week. I felt like my good luck just kept coming.

We were made to stand at attention behind our chairs, and we had to stand for a long time. The cadre milled about and ganged up on whichever RAT they could catch glancing timidly around. When the order to sit was finally given, we all took to our chairs as fast as we could.

I was starving, so I was ready for a hearty meal. That was far from what was about to happen, but it still had not sunk in all the way that I was in the wrong place. Just maybe my dad had known a bit more than I had given him credit for.

"Sit at a brace, maggots," Sgt. Swan yelled. The noise of cadre yelling from every table was like a roar.

"What's that?" Cadet-Recruit Bose asked as he turned his head to look at Sgt. Swan.

Bose sat directly across from me, so I was destined to look at him while he looked at me as we ate for the next semester. I don't remember him as an attractive fellow in any regard. His bulging eyes were made more so behind his thick glasses, and his chin was too far from his forehead for the width of his face. I didn't know it at the time, but when his hair grew back, it was so blond that it looked white. I am certain that his memory of me lacks any remembrance of my youthful beauty either.

"WHAT'S THAT, SERGEANT?!!!" Sgt. Swan yelled back. "Eyes

straight ahead you idiot. Why in the world are you looking at me? Are you crazy? Did you forget to bring your brain with you? Don't you ever look at me or it may be the last thing you ever see, you dumb RAT. I can't believe they would let a raging homosexual like you into the corps. What is this world coming to?"

Bose looked back at me, and I could see his Adam's apple bobbing up and down like a puppy trying to escape from under a blanket as he swallowed a bit of shock. It was a big Adam's apple.

"I said to sit at a brace," Sgt. Swan repeated. "Look over here you idiot RATs."

No one turned to look. We had been severely coached to not look mere seconds ago, and now he was telling us to look. It was confusing, distressing, and upsetting all at once. At least we weren't alone. From all over the mess hall, the yells of old cadets at the nervous RATs continued.

"LOOK OVER HERE NOW!" Sgt. Swan yelled.

We all turned toward the angry sergeant.

"When you sit at a brace, fist distance from the back of your chair, your ribcage is on the table, your eyes are straight ahead, your shoulders are back, and your chest is out," Sgt. Swan said as he demonstrated the position. "It is like standing at attention while you are sitting down."

He demonstrated the position easily from his RAT year of practice.

"You will eat a square meal. Lift the food from the table till your fork is level with your mouth. Bring the utensil directly to your mouth and deposit that food. Move the fork straight back out and lower it to the plate. Once you have placed the fork on your plate and removed your hand from sight, you may chew your food for four seconds before swallowing. Do not chew while the fork is in the air, and don't you let me catch you with more than a four-second chew time. You got that, maggots?" he asked. "Now commence. Eyes front and sit at a brace."

We turned back toward the center of the table and did our best to sit at a brace. It was amazingly uncomfortable and inconvenient, yet as I reflect on it, it is surprising how normal it felt by the end of my RAT year. It still feels normal today when I am telling this story to friends around a table and demonstrating the position for effect.

To eat, a new cadet had to request permission. I still remember as if it were yesterday learning the method to request anything as a RAT. It was done by placing your right fist out in front of you and saying, "Excuse me, Sergeant (or whatever was the highest rank present), Cadet-Recruit Gregory, C.L., requests permission to use my hands and utensils for the preparation, transportation, and consumption of my food and drink."

You would then sit there at a brace with your arm out until the ranking old cadet decided if you were to go on with your request or not. This same method was used to request just about anything that a normal human gets to do just because they are a normal human, for example, "Excuse me, sir [or, rank of the person I was addressing], Cadet-Recruit Gregory, C. L., requests permission to (stand, sit, walk, run, drink, leave, throw something in the trash, go to the bathroom, put on my hat, take off my hat, scratch my nose, etc.), sir." The old cadet then would decide the RAT's next action.

The fat kids had it bad. They were not allowed to eat until there was almost no time left for chow. However, it was not that long till there weren't that many fat kids to be found. Bad-tasting food, high-speed PT, and a policy that made it hard to get enough calories to live, much less gain weight, made fat kids a rarity. Being that it was a 1983 version of America helped that fact, as well.

That first week, there were major infractions as we learned how to eat as a RAT. If you touched the table with your hand, you had to "mount the twins," (put your hand over the salt and pepper shakers) so that another cadet could smack the back of your fingers with the handle of a butter knife. That may sound like a small thing, but it could cause the kind of pain that would last for days.

When pouring beverages into our glasses, we had to hold the glass against our chest and pour with the pitcher in such a position that any mistake would end up on our food or uniform. Since this was done from the brace position, it was necessary to keep your finger in the top of the glass to know when it was full. It is amazing how hard that is to do, yet how quickly you learned the skill when you had to eat red, fruit-punch flavored, powdered mashed potatoes a time or two and had the stain on your uniform to advertise your inadequacy.

When the meal was done, each new cadet would put his right hand forward and sound off with, "Excuse me, Sergeant. Cadet-Recruit ___ requests permission to take your tray and mine to the Den of Culinary Atrocities." This ensured that the new cadets would be clearing the table.

With our meal coming to an end, none of the RATs had succeeded in getting more than a few bites without being harassed and penalized. The order was finally given to "grub up," and that gave us a few seconds of diving into our food without regard to hygiene, manners, or custom. It was literally the last five seconds of a meal, and each RAT would try to shove what was left of his meal into whatever size mouth God had graced him with. Luckily, I had been graced with a good capacity and a fast-swallowing mechanism that came to serve me well later in the year when I was named the Delta Troop Jell-O-Slurping Champion. I did advance in the Jell-O competition circuit but got beat at battalion level by a RAT in Echo Troop.

On the heels of the grub-up command came the order to "fall in." This meant that we were to be outside and in formation. While we still didn't know that or where we fit into the scheme of the platoon and company, we were herded and prodded like lost sheep by our angry shepherds as we tried in vain to make the old cadets happy. Once formed up, Delta Troop was marched back to The Box and our indoctrination continued.

Since we had just had lunch, we weren't run or physically worked immediately upon getting back to The Box. 1st Sgt. Wengert had the entire group of new cadets sit on the grass that was between Delta Troop Walk and the HQ, or Headquarters, Troop Walk.

"Sit down a spell, boys," the tiny first sergeant said. His calm tone and easy manner implied that we were about to have a comfortable introduction to our first sergeant.

It was with great relief that we all sat down on the cool grass. The summer sun was hot, but there was enough of a breeze and enough sweat on each of us that I actually remember this brief moment with fondness.

"I guess you are all a little nervous," 1st Sgt. Wengert continued with a knowing smile and slight chuckle. "Don't worry, boys. I was there myself once upon a time back when this corps was really tough."

"When was that?" RAT Jones asked.

"WHEN WAS THAT, FIRST SERGEANT, YOU PIECE OF CRAP MAGGOT WITH ROTTING DOG%$&* FOR BRAINS!!!" 1st Sgt. Wengert yelled as he spun on RAT Jones and ran at the cowering new cadet. "I am not your #$%^ing friend, counselor, preacher, father, big brother, or comrade. I am your worst $%^&ing nightmare. Knock out some push-ups for being so &^$&*%*& stupid, and never address me again in your life without my rank. Do you want to be the first RAT to #*^%ing die in Delta Troop this year?"

It was frightening. From kindly and warm to psycho-murderer has rarely been displayed with such ferocity and speed. Cadet-Recruit Jones was trembling and crying as he did his push-ups with 1st Sgt. Wengert leaned over and cussed in his ear. For the rest of us RATs, we were sitting up at attention as we tried not to look at Jones but couldn't look away. The cool grass and peaceful moment had passed much too quickly.

With that episode over, it was becoming clearer to all of us that we were at no time to trust the supposed kindness of our cadre. It didn't come up too often when you would think that one was being nice, but when it did, it invariably led to a scene like the one just witnessed.

Capt. Sheen stood up and gave his introduction. He had just returned from US Army Ranger school, and he was a very hard and disciplined cadet for being a sophomore in college. His introduction was short and to the point, and it was obvious that he had other things on his mind besides the bunch of useless and grody RATs that sat in front of him. He did mention that 1st Sgt. Wengert was in charge and his word was as the word of God.

Each platoon leader, platoon sergeant, and squad leader then stood up and made a short introduction. These were brief and to the point, but each one wanted to be the next battalion commander or sergeant major. I did not take notes during these introductions, but I did enjoy the free moments without harassment. To my left, Milt, my roommate, had fallen asleep. When the world crashed down on him, it was something to behold, and it was perhaps the fifth time in one afternoon that I had seen one of my fellow RATs break down and cry. Sleeping at a time like this was apparently neither advisable nor acceptable.

With the hazing of my roommate complete, the squad leaders gave a short pencil, like the ones you get on a golf course, a piece of NMMI stationery, and an NMMI envelope to each RAT.

In a bellowing voice that was in stark contrast to his size, 1st Sgt. Wengert said, "All right, maggots. Write what I am about to say. 'Dear Mom and Dad.' If your Daddy left because you are such a piece of &$%t, just put, 'Mom.' If she couldn't stand you either, write 'Grandma.' If that doesn't fit your &*^%ing foster-child life, put in the title of the sorry sack of ^%*t that is responsible for bringing you here to make my life miserable."

Of course, we didn't write all of that, and mine was luckily the simple 'Dear Mom and Dad.'

"On the next line, start your first letter home with these words. Write, 'I love it here. The food is great, and all the cadets are really nice to me. Thank you so much for sending me to NMMI.'" 1st Sgt. Wengert continued, "Now, sign your name. Place the letter in the envelope, seal it, and address it home, and it will be on its way tomorrow. But it won't beat some of you quitter maggots home, mark my words. The US Post Office is slower than a cowardly RAT running with its tail between its legs."

Of that, he was decidedly right as we lost about five more guys before that horrible day was over.

Each squad leader then took his RATs to a separate area for

continued teaching, using the word "teaching" here in its broadest sense. All of us were learning with a fervor that only pure fear can provide. The fear made learning fast despite the limited ability of our teachers to teach and our exhausted minds to comprehend.

For the next several hours, we were taught how to stand at attention, stand at parade rest, and do right-face and left-face moves. We learned how to form up, where to keep our eyes fixed, dress-right-dress, and more than anything else, how to do push-ups. When one of us would mess up, Sgt. Swan would have that individual sit down in a chair he had provided for that purpose, and the rest of the squad would do push-ups while the offending RAT watched. While that is a very effective training technique by itself, the later RAT-to-RAT hazing is what makes the offender start to comply. You certainly did not want to make your fellow RATs hate you when you already had every old cadet as an enemy.

Late that afternoon, we were dismissed to our rooms with orders to undress, put on our bathrobes, and bring a towel.

"Fall out to your rooms you mommas-boys, get in your robes, and grab a towel," ordered 1st Sgt. Wengert.

We sprinted up the steps with what remaining energy we could muster, and each donned the red NMMI bathrobe that we had been issued and grabbed a new white towel to go with it. Shoving our feet into flip flops, we ran back to the troop walk as fast as our footwear allowed and formed up again.

1st Sgt. Wengert was not impressed with our speed, so he made us go back to our rooms and return to formation again as a practice. Then again.

The steps to the top stoop, which was the living area of high school RATs, were metal. The concrete of the stoops was polished from the action of thousands of cadets through the decades, and flip-flops are uncommonly hard to run and navigate in. More than one of us fell in the process.

Finally, convinced that even practice would leave us wanting, 1st Sgt. Wengert decided it was time for us to shower.

In minutes, the entire Delta Troop of RAT cadets formed up in a sea of red bathrobes. Although I was happy that it looked like it might be shower time, I was totally unprepared for "airborne showers."

Sgt. 1st Class Franco gave the order of "right face" to our platoon. We all swiveled to the right like an unoiled machine, but at least we turned to the right for the most part.

"Single file, forward march," Sgt. 1st Class Franco bellowed, and 1st Squad began to march. As the last man of 1st Squad passed the first man of 2nd Squad, 2nd Squad moved out. It was repeated from squad to squad until the entire platoon was marching in a single-file line.

"Put your arms out like a plane and make some engine noises, maggots," Sgt. 1st Class Franco ordered. "Double time, march," he continued as we complied.

So now there is a platoon of about 30 RATs that are naked under red bathrobes, some tied, some untied, running around in a single file line with our arms spread wide and making noises like plane engines. We ran up and down The Box, returned to Delta Troop walk, and followed the 1st Squad leader, Sgt. Hoover, up the stairs to the third-floor sinks. All the troop bathrooms were referred to as sinks, even though we had an actual sink in our rooms, so the "sinks" were just toilets and showers.

There was a common shower area with five shower heads in a row on one wall. The cadre had turned all the showers on, with the first one being completely cold, the second hot, then cold, then hot, then cold. We were lined up outside the door of the sinks, still making our plane noises.

"HOOK UP," came the order from one of the cadre.

This meant that our right arms were bent, and we were pretending that we were hooked to a static line in preparation for jumping out the door of a plane with a chute on our backs. The first RAT at the door of the shower was ordered, "Jump." At that order, he dropped his robe and jumped into the first cold shower.

Every ten seconds, the order to jump was given, and the RATS in the shower would move over as the next one entered. This gave each of us a fifty-second cold-hot-cold-hot-cold shower. Never mind the soap. Airborne showers are not necessarily the best way to clean a body, but you can certainly rinse a lot of them off fast that way. For the first six days, this was to be the only way that we would get to shower. But, we were good at it by the end of that week. Plus, there were fewer of us by the time Hell Week had reached its end.

29

EATING RATS

The rules for eating were strict and were not easily taught, understood, or obeyed. To find out what the rules were, a RAT had only to do anything that normal people do normally one time, and it would be pointed out that this was not the way to do things if you were a RAT, and therefore, abnormal.

By the first morning, we had had the experience of two meals at a brace. All body parts were sore from the amount of physical abuse we had endured from countless push-ups and endless running. Cadet-Recruit Bose stared at me across the table as I stared at him.

"What do you think of Cadet-Recruit Gregory, Mr. Bose?" Sgt. Swan asked.

"Nothing, sir," Bose answered.

"How many times do I have to tell you that I am not a sir? You owe me twenty when we get out of here," Sgt. Swan returned. "Mr. Gregory, what do you think of Bose?"

"Nothing, Sergeant," I said, proud of the fact that I was such a quick study and did not call him sir.

"I think Gregory is a piece of crap," Sgt. Swan said to Bose. "You do too, don't you Mr. Bose?"

"I don't know, Sergeant," Bose said.

In fact, neither of us had spoken a word to the other since Bose was a college RAT and lived on the first stoop. We had obviously seen each other plenty over the last 18 hours or so but had not actually spoken.

"What do you mean you don't know?" Sgt. Swan asked. "I just told you what you think. You think Gregory is a piece of crap. Whatever I tell you to think, that is what you are to think. Understand?"

"Yes sir," Bose said, and I saw the grimace he made as he realized an instant too late that Sgt. Swan had not been promoted to an officer in the last few seconds.

"I am not a SIR! You are an absolute eye-dee-ten-tee. I work for a living, and you owe me fifty," Sgt. Swan said. "Now, what do you think of Gregory?"

"He is a piece of crap, Sergeant," Bose said.

"Good, maybe you can be taught," Sgt. Swan said. "Now, I want you to kick the crap out of Gregory."

"What, Sergeant?" Bose asked.

"Kick Gregory as hard as you can," Sgt. Swan said.

Bose looked at me, and I am certain that I had a wide-eyed look on my face in the anticipation of the pain.

"What what?" Sgt. Swan asked. "Did you not hear me or not understand me? I want you to kick Gregory in the shin as hard as you can."

"Yes, Sergeant," Bose said. He looked at me with an apologetic look, cocked his leg, swung as I flinched, and there was a loud thunk, the table vibrated, and Bose yelped.

Down the center of the tables at NMMI there was a supporting board that was situated about twelve or so inches off the floor. It was impossible to kick the man across from you since this board would block the blow and hurt the kicker. It was an often-used stunt, and later, it became a punishment when all the RATs knew about it but were told to kick the man across from them anyway. How much the table moved was an important part of determining if the punishment had been successful. Not enough movement meant another kick was required.

Tears formed in Bose's eyes, but he sat there and did his best to absorb the pain. I felt sorry for him, even though I was happy I had not ended up on the end of his punt.

It is officially against the rules to physically haze a RAT, but that doesn't mean that they can't physically haze themselves. An argument could be made that when an old cadet forces a RAT to do it, it is physical hazing anyway.

Supper Roll Call, or SRC, was a memorable night on the third night of Hell Week. We were dressed in our Class B uniforms, which consisted of a long-sleeved khaki shirt with full brass, green pants, low-quarter shoes, a garrison cap, and a tie. None of us knew quite how to put the uniform together yet, but we did the best we could.

During Hell Week, many of the meals were eaten in PT uniforms. We were being run, marched, drilled, and otherwise put through our paces, so we did not have the assembly and use of our uniforms down to the point that we could be seen in them. Learning how to put all of it together was a part of Hell Week, and this was our first meal in our Class B uniforms.

"What in the #%&* is your name tag doing on your shirt so crooked?" Sgt. Swan asked me. Although he was only an inch or two from my face, he was not using his inside voice, nor could he see my offending name tag from that position.

"I don't know," I answered honestly.

"I DON'T KNOW, SERGEANT!" He yelled back. "You must be the dumbest RAT that has ever put on a uniform. Drop and give me twenty."

I went down to do my push-ups, but there was not much left, and doing twenty was out of the question. I quivered and shook while I tried, and RATS all around me were being dropped for similar disgusting behavior. Before long, Sgt. Swan and Sgt. Johnson had the whole squad on the ground. They were just getting back to the first one dropped, me, and squatting down by my head as they discussed my weakness.

"Which maggot is this?" Sgt. Johnson asked Sgt. Swan as they crouched in front of me.

"This is one of the dumbest ones," Sgt. Swan replied. "He thinks he is here to rodeo, but he is really here because his mom doesn't know who his daddy is and is tired of trying to figure it out by looking at him."

"Sounds about right," Sgt. Johnson said. "Hard to believe that he was ever the fastest swimmer."

That was a common putdown, making fun of the fact that each of us had been the result of the strongest and fastest sperm.

"Must have been a pretty small selection of pathetic dog-paddlers in that batch. What are we going to do with him?" Sgt. Johnson continued.

"If he don't do some push-ups pretty quickly, I'm going to drop him out his third-floor window tonight," Sgt. Swan said. "That way we won't have to deal with him and the whole corps will be a better place."

I did my best to raise my body from the ground, but there just weren't any push-ups left in me. Sgt. Swan leaned down and whispered quietly in my ear, "I am going to %*^&ing kill you, maggot, for being such a worthless piece of crap. Don't you even think that I am not deadly serious."

In hindsight, I know that there was really no way that I was going to be murdered. But, in moments like that, one had to wonder.

At one meal in the early days, I leaned slightly toward my fork as the food was coming toward it. Sgt. Swan noticed immediately.

"Cadet Gregory, that is the nastiest thing I have ever witnessed," Sgt. Swan said. "The difference between a man and a dog is that a man takes his food to his mouth, and a dog takes his mouth to his food. If you want to eat like a dog, let me help you."

With that, he placed my tray on the floor and I had to finish my meal on all fours. As I remember it, at least I had a pretty good meal that night. I guess good is not the right word. Let me say it was at least plentiful. Old cadets ate with fewer manners than a dog surrounded by his litter mates, so he had in fact witnessed and been part of taking his mouth to his food, but that was not something that any new cadet could point out to an old cadet.

As the year wore on, we were introduced to some of the eating games that are such a hit. One was called, "pass the spoon." In this game, the handle of a spoon is held between the index and middle finger and passed from one cadet to the next. Each cadet would then add a food item of his choice to the spoon until it was so full, or clumsily held, that something would spill. Whoever had hold of the spoon for the spill would have to eat whatever was on the spoon at that time.

Of course, it was often a most disgusting mix, especially when you consider the material that we had to start with. Most of the food from the mess hall was barely edible in the first place and was much worse when it was mixed with the intention of making it as gross as possible.

It was not uncommon for an old cadet to add something like a heaping helping of pepper or salt, and on unlucky occasions, a pinch of tobacco. Of course, they added to the spoon while the new cadet closest to them held it. Mix that with a little milk, a piece of mystery meat or elephant scabs, some Jell-O, a pea or two, some butter, and perhaps a dab of unidentified frosting from the desert that night. On the bright side, at least the loser was getting to eat.

Another popular spectator sport was the Jell-O-slurping championships. Each squad would have practices when there was Jell-O, which was at least every other day. The trick to being good at Jell-O-slurping is to time the swallowing perfectly with the breathing. Getting the timing off could result in Jell-O going down the wrong pipe, and that is not a great experience. It is a funny experience for everyone who gets to watch the unfortunate would-be champion spewing red Jell-O out his nostrils, though.

The practice gained in the 2nd Squad, 2nd Platoon, Delta Troop really paid off for me. In no time, I was easily winning the local heats at Sgt. Swan's table. This led to platoon and troop matches, and I was the Delta Troop Jell-O-slurping champion the first semester at NMMI. I eventually lost to a RAT from Echo Troop, so I didn't ever get past the battalion championships. I think I could have with the right practice and coaching.

Other new-cadet games that were popular with the old cadets were things like Odd-or-Even. This was a simple game where an old cadet would grasp your arm or leg hair and yank out as many as came with the grab. You would bet whether there was an odd number or even number of hairs. Losing meant lots of push-ups, but, the new cadet had already lost when the old cadet pulled out the hair to start with. What fun was had by all.

The food at NMMI was atrocious, but I have learned that perspective has a big part to play in one's appreciation for available cuisine, and I had come from a place where both my mom and grandma were amazing cooks. There were those amongst us who thought the mess hall was just fine. That thought was beyond me.

We often had a round patty of meat product that was referred to as "elephant scabs." It was brownish-gray in color, and the outside edges would be tough and dry, while the middle texture was slimy and wet. It was covered with a red sauce of sorts that would also change color and texture from the edges to the middle, going from gummy and darkish brown, (think dried blood), to thin and wet, (think fresh blood). The flavor was not bad for the most part just inside the tough edges but outside the slippery middle, so it was possible to enjoy about 50% of the offering. Couple it with powdered mashed potatoes, gray beans that should have been green, and enough salt and pepper to mask it, and you could at least sustain yourself.

I had been raised in a home where chicken was served less than once a year. For a few years of Dad's youth, his miner dad had added a chicken-raising operation to the inventory of money-making endeavors, so Dad hated chickens. No, hate is not too strong a word. Cholla made amazing fried chicken, made more amazing by the fact that it was almost never served. My mom's Hispanic heritage and Mexican food cooking skills coupled with my dad's required dosage of red chili to make up 75% of his blood meant that fried chicken was as rare as caviar in our household. I never tasted caviar as a kid.

At NMMI, fried chicken was served at least thrice a week, but looking back, I think it may have been eight nights a week. Man, was

I sick of chicken when I finally left that place. It was produced in great vats, with the bottom pieces being submerged in oil for generations, while the top pieces were scorched and dry enough to be considered jerky. There was a middle batch that was old-cadet chicken, so I was only exposed to that during the occasional lucky lunch and the two years following my RAT year when I was an old cadet, enjoying all the privileges that were included.

Lunch was the only meal not eaten with the squad, so it was the meal that most new cadets were able to get the calories they needed. Another potential way to get a decent piece of chicken was when you happened to be on mess hall duty or had a friend who was so inflicted. You might get a piece of the old-cadet chicken, which was tolerable.

Breakfast was powdered eggs, which, like the chicken, varied in flavor greatly depending on the position they held in the pan they were prepared in. This went for the sausage links and bacon as well. You could get eggs that were about to return to powder or those that might be considered powdered-egg soup. Luck and connections to the server played another huge role in this, and it wasn't until you were an old cadet that you were able to enjoy the 10% of the eggs that looked like the picture on the powdered egg box. I never saw the box, but I do imagine a picture of a 1950s apron-wearing blond mom, smiling over the plate of powdered eggs in one hand, spatula in the other.

The only thing that the cuisine at NMMI had going for it was the fact that one was always hungry. I had the sort of appetite that almost defined me. Someone might refer to a redhead as, "you know, Tom, the red-headed kid," or call the tall guy, "John, the tall kid." With me, it was often, "you know, Chris, the kid who eats four times more than anyone else." With that sort of appetite, I had to eat, so I learned to get it down even if I didn't like it. Besides, there is no spice like hunger.

Once I left the confines of the Institute, I was a picky eater for many years. Age and travel have expanded the list of foods that I will eat, but for at least a decade after leaving NMMI, I wouldn't consume anything that I didn't want to.

The experience of military school also taught me to eat at a very high speed. Even today, I am often done well ahead of anyone else at the table unless I concentrate on taking my time. Not a good habit for health, but some habits get burned-in deep.

30

RAT RAID

Hell Week was behind us, and the regular life at NMMI had begun. We still formed up for BRC and SRC (breakfast roll call and supper roll call), so we had two meals a day at a brace. Lunch was on our own time because everyone had different class schedules. We also got to eat Sunday morning breakfast on our own, but surprisingly, a lot of cadets would choose to sleep instead of eat. Not me.

"How's it going in 1st Platoon, Jack?" I asked one of my fellow Delta Troop new cadets. I was laying on my bottom bunk. He was sitting on the shine box, which was a hard metal cabinet about a foot wide and eighteen inches high, filled with the stuff we used to shine our shoes.

"My squad leader is an SOB," Jack answered.

"They're all SOBs," Pips said. Pips was also from Delta Troop and sitting on one of the chairs. Each room had two chairs, two bunks, and two shine boxes.

"I've got a buddy in the Pixies," Jack said, referring to HQ Troop, which was the band. "He said the old cadets there are all pretty nice."

"All the Pixies do is beat and blow," Bill said. Bill was sitting in the other chair, enjoying a huge dip of snuff, and offering rude comments as needed.

My roommate, Milt, was lying on his top bunk. He was universally disliked. A grode from the word go, Milt had a bad attitude that ended up getting those around him in trouble as well. He was late, sloppy, unfriendly, and had a constant look of insolence. That he did not want to be at NMMI was evident in every action. I was able to get a roommate change after we came to blows a few times, but this was before that happened.

"I'm just looking forward to our first rodeo," I said. "Three more weeks and we will be..."

I didn't get to finish because the door slammed open. The two old cadets that had found us lazing away a Sunday afternoon were about to have some fun.

When the door to the room was slammed open, in came Staff Sgt. Bach and Cpl. Morris. "RAT RAID!" Staff Sgt. Bach yelled.

Staff Sgt. Bach and Cpl. Morris were both linemen on the football team, and the little room that was crowded before was now bursting at the seams.

We RATs were all moving to "hit a wall," which was the required RAT action when an old cadet entered one of our rooms. We were not allowed to lock our doors when we were in our rooms, ever, so old cadets could bust in anytime and haze us. There was also a rule against five or more RATs in any one room at a time. None of us ever counted Milt as one of the five, so we were caught.

The four of us who were motivated had hit a wall at speed, slamming our backs into an open area and coming to attention to await the malicious demands of our superiors. Milt had slowly swung from his bunk and stood against the closed locker door, neither slamming his back nor holding a very good model of attention.

"I count five eye-dee-ten-tees here, myself," Staff Sgt. Bach said to Cpl. Morris. "What do you make of it?"

Eye-dee-ten-tee stood for idiot, just like bee-one-are-dee stood for bird.

"You couldn't be more right about that one, Staff Sergeant," Cpl.

Morris said, his voice taking on the qualities of a tent-revival preacher. "Looks like plottin' and conspirin' in our midst. We must cure this cancer afore it spreads to the troop and to the corps."

"Amen, Brother Corporal," Staff Sgt. Bach said. He had been soaking a washcloth in the sink, and he threw it on the floor. When it hit the ground, he hollered, "GRENADE."

Every one of us made a dive toward the wet washcloth in an effort to "save our buddies" by dampening the explosion of the pretend grenade with our bodies. Milt moved slower than any of us, so he barely made it to the pile before Cpl. Morris yelled, "SHARK ATTACK." None of us ever doubted that we would all be dead if Milt was the one needed to save us from an actual grenade.

The warning of a shark attack meant everyone had to get completely off the floor. It was like a mad game of Twister, and you might find one leg on a chair and another on a shine box. Maybe you would have to hang from the locker door. It didn't matter, you had to be completely off the floor.

About the time everyone was up, the resoaked washcloth had become a grenade again and the sharks were gone. With that grenade smothered, the sharks reappeared. This would happen for as long as the old cadets found it funny, and the damage it could do to a clean and orderly room had to be seen to be believed.

I had found myself on the Milt's top bunk after the last shark attack, and he was alone on the desk closest to the door. Cpl. Morris threw the wet washcloth close to Milt and yelled "GRENADE," but I still beat him to cover it.

"I have the opinion that the fat one is too slow to save his buddies," Cpl. Morris said, referring to Milt. "Although he is certainly endowed with enough around his middle to confine the explosion of even the most determined grenade."

"Yes, I do believe you are correct in that assumption," Staff Sgt. Bach answered. "His t-shirt has yet to even become damp from the blood of his devoted followers. Perhaps he would enjoy watching them work for him?"

The rest of us were soaked from jumping on the "grenade." Milt had yet to have much more than the odd droplet hit him.

"You have never been more right, my esteemed staff sergeant," Cpl. Morris said. "FALL OUT TO THE FIRST STOOP, MAGGOTS." He yelled, going from his fake genteel speech with Staff Sgt. Bach to the demeanor of a drill sergeant in less than a breath.

I was in flip-flops, Milt was barefoot, but at least Bill, Jack, and Pips were wearing tennis shoes since they had been visiting. We stood at attention on the bottom stoop as the old cadets yelled in our faces and stomped around us like madmen, focusing on the four of us but leaving Milt out of the hazing.

"I think fatso should take a break," Cpl. Morris said to Staff Sgt. Bach. "We do not want to damage his delicate sensibilities."

"Yes, yes, but of course you are correct," Staff Sgt. Bach said. He retrieved a chair from one of the rooms on the first stoop and had Milt sit in it behind us. He was facing the middle grass in front of our troop; we were all standing at attention facing the rooms with our backs to Milt and the two old cadets.

"Are you comfortable, old son?" he asked Milt.

"Yes sir," Milt said, which was as far from the truth as you could get.

"Now now with that sir stuff. Just call me Staff Sergeant," he patted Milt's shoulder.

We were all learning that there are few things as mean as a nice old cadet. This was part of that training.

"Why don't you say, 'about face' to your slaves?" Staff Sgt. Bach said.

"I don't have any slaves," Milt said.

You could tell the Staff Sgt. Bach was about to the end of his rope with Milt, but the fun was still coming, so he pardoned the denial as well as Milt leaving out addressing him as, "Staff Sergeant," at the end of it.

"Well, say the words anyway," Cpl. Morris said. "Say 'about face' in a clear crisp voice if'n you don't mind."

Milt said, "About face."

None of us moved because we didn't yet know what was happening or that the quiet words Milt had said were for us.

Both old cadets came alive. They were in our faces, screaming at us, "DIDN'T YOU HEAR YOUR MASTER? HE TOLD YOU ABOUT FACE, NOW ABOUT FACE WHILE YOU STILL HAVE FACES. ARE YOU GUYS THAT STUPID?"

There was of course more, but you get the gist. We turned and stood at attention, all of us now facing the grass and standing just behind the seated Milt. Since this was our first round with this game, none of us knew what was coming.

"Have 'em run, old son," Cpl. Morris said to Milt.

Milt held still and quiet.

"You better say, 'run,'" Staff Sgt. Bach said through his closed teeth. Still quiet, but peril was promised.

"Run," Milt said.

We knew it was for us, so we all took off across the grass. Flip flops are not great for this as I instantly recalled the airborne showers of a few weeks prior.

"Stop," Milt yelled as we had made several yards across the lawn. "Drop," was his next command, so we all went into a front-leaning rest.

"Run," he sounded off. "Stop." We could barely hear it. "Drop," so we did. "Roll right," which we did. "Run, stop, about-face, drop, roll left, run." Again, we did all that we were told.

Our wet from the grenade-explosion shirts were now getting muddy from the outside and wetter from the inside as we sweated from the workout. Commands were coming from an uncomfortable Milt, and the old cadets looked far from bored.

All good things must come to an end, and thankfully, that rule stands for bad things as well. We were finally released to our rooms, and Milt and I returned to the ruins of our once neat and tidy room. I went straight to the shower telling Milt that he better start cleaning up. When I got back, the room looked the same and my eye-dee-ten-tee roommate was in his bunk. This might have been the catalyst for our first fistfight, but I don't remember for sure. We had a couple, and I had a new roommate by the end of week six.

Talk of a "blanket party" for Milt was circulating amongst the new cadets. This is done by pinning a sleeping cadet under his blanket at night and whipping him for his misdeeds. You might hear someone talk about "locks-in-socks" or talk about putting a bar of soap in a sock as a weapon, but the only blanket party I ever knew of that really happened was just fists and elbows. It was enough.

Milt was not given a blanket party, and thankfully he was gone after Christmas. Whether it was his choice to leave or whether he got expelled, I don't know. I just know that the corps has a way of eliminating those who are found wanting, and few fit that category better than Milt.

31

MY PERSONAL TIPPING POINT

From an academic side, NMMI was so much tougher than any school I had yet experienced. My mom had taught me to read well ahead of kindergarten, so my early public school years had not only been academically easy but extremely boring. I alluded to being a difficult youngster, and part of that was the boredom of being taught stuff that I had already learned, and that was being taught at a snail's pace. Well, NMMI solved that problem for me.

There was a mandatory two and a half hours of study required of every cadet, five nights a week. You were to be in your room, at your desk, and working on schoolwork. Tactical (TAC) officers would make the rounds to ensure that you were doing as directed. If you were caught in your bunk or doing stuff besides schoolwork during that time, you would "get a stick," meaning written up, and assigned "tours." This basically meant that you owed a certain number of hours walking around with your weapon going nowhere.

To be honest, when a TAC officer was on the stoop, pretty much everyone would know about it, so getting tour duty for misbehavior during study hall was rare. People would knock on their walls to warn the next room about a TAC officer, and the warning would be passed on, just like flashing your lights at oncoming traffic when you know a cop is hiding for a speed trap.

You could end up on tour squad for a lot of different reasons, which is a fact that I proved more than once. Missing a class meant a ten-tour stick; ten hours of marching around with your rifle on your shoulder for missing a one-hour class. That was not a good trade. You could get a stick for a sloppy uniform, walking in the wrong place, disrespect, being tardy for an appointment, or a slew of other things. The biggest was getting caught drinking. You could end up in a "pickle suit," which was the olive-drab fatigues that we all had to wear when doing army field operations and were the main Army uniform in Vietnam. The camouflage BDU (battle dress uniform) was new for the Army at that time, so a lot of the gear we used was left over from Vietnam or before.

Our rifles were to be considered almost a holy relic. When asked what our rifles were, we had to say the following, word for word. I still remember like it was drilled into my brain and branded there yesterday.

"Sir, my rifle is a US Rifle, Caliber 30, M1903, lightweight, manually operated, bolt-action, breach-loading, shoulder weapon. 8.69 pounds, four-three inches in length, sir!"

The rifles we had at NMMI had been rendered inoperable, but I am certain that some of them saw service in WW II, Korea, and Vietnam. They were cumbersome and beautiful at the same time, but mainly cumbersome remains in my mind. As I later treated my M-16 when I was in the actual Army, our rifles were cared for as if they were newborns.

"Cadet-Recruit Gregory, would you like to come down to my room? You can hang out and listen to some music," Sgt. Swan said.

Had I heard that right? An old cadet offering to hang out? "Yes sir," I answered. I knew his room well. I had been making the bed and cleaning that room every morning for a month since Hell Week had ended.

"Now now, Gregory. You know I'm not a sir," Sgt. Swan laughed. "You really need to work on that whole sir vs. sergeant thing," he patted me on the shoulder. "Why don't you come down about 1600 hrs. That'll give us a couple of hours before SRC, and you can hang out with us."

I was happy. I was finally getting a chance to make some friends with power. My once sadistic squad leader had suddenly changed. Maybe this would be the 'in' that I needed to get a little relief. Someone in charge who knew humans shouldn't treat other humans this rough, even if that other human was a RAT. I had not expected this from Sgt. Swan, or any other old cadet, for that matter. Their minds seemed more interested in torture techniques than compassion.

I showed up at 1600 hours, ready for some real soldier bonding time; to get to know each other on that basic level so maybe my life would get easier.

"Hey, Gregory," Sgt. Swan said when I knocked on his door. "Where's your rifle?"

"My rifle?" I asked.

"My rifle, Sergeant," he corrected me.

"I'm sorry, Sergeant," I said. "I didn't know you wanted me to bring my rifle."

"Oh yeah," he said. "Go get the ol' girl. I'll bet your rifle will like the music as much as you do."

"OK," I said, doing an about-face and heading back to my room to get my rifle.

"OK, Sergeant," Sgt. Swan said. "We'll need to work on that. Hurry up now."

What in the world? Why did he want me to bring my rifle? My Spidey sense was tingling, but I hadn't figured out what was going on yet. But man, was I in for an afternoon.

"Come on in," Sgt. Swan said when I returned. "What kind of music do you like?"

"Country," I answered.

"IF YOU FORGET TO CALL ME SERGEANT ONE MORE TIME, THEY WILL NEVER FIND YOUR #^*^!&* BODY. NOW LOCK IT UP, MAGGOT."

Sgt. Swan was in my face, and I was standing at attention. I could see that this was not going the way I thought it was going to go. My team leader for A Team in the 2nd Squad was Cpl. Smith. He had been invited over and was sitting on the bottom bunk in his PT uniform with a smirk on his face that offered a hint at the torture I was about to endure. There was another old cadet at the desk, and I didn't know his name at the time, just that he was big, blond, and had a face that promised the brain behind it was dim.

"Grab the top of the locker door, maggot," Sgt. Swan commanded.

I did as I was told. Luckily, I was tall enough to do this without having to jump for it.

"Now, bend your knees."

I did this, too, so my weight was now hanging from the locker door. He placed my rifle across my ankles, then went to his stereo and turned up some of the worst music I had ever heard. I think it was called electronic rock, or something like that. It was nothing like music, in my humble opinion, but at that moment, the assault on my ears was the smallest of my problems.

Pretty soon, the weight of the rifle across my ankles and the top of the door biting into my hands was starting to tell. It seemed like it had been at least half an hour, but the truth was probably more like half a minute.

"Don't you let that rifle hit the ground, maggot," Sgt. Swan said, the malice evident in every syllable. "How you likin' these tunes?"

I hung in silence, trying to just stay on the door and keep my rifle off the ground.

"Didn't you hear me, Gregory?" Sgt. Swan asked, getting closer to me. "HOW DO YOU LIKE THESE TUNES?"

"I don't like them," I answered.

"I DON'T LIKE THEM, SERGEANT!" he yelled, standing beside and behind my hanging form, his spittle joining the sweat on the back of my neck.

"If you don't like this music, maggot, you may need to listen to it longer to gain an appreciation of the intricacies that make it so beautiful. What do you think?"

I didn't answer. What kind of answer could I have given? The fact of the matter was that there was not a good answer to be had. If I had agreed, then they would haze me for not standing up for my opinion of the music. If I had disagreed, they would haze me for refusing to join in their opinion. This was a common thing to happen to every new cadet where the situation was set up so there was no winning. My choice of not answering was also wrong, as I was about to find out.

"I asked you a question, Mr. Gregory. Did you not hear me or are you too stupid to understand the question?"

"I don't know, Sergeant."

"What don't you know? There are now several questions for you to answer. Do you think you need more time to appreciate the music, how is your hearing, and are you too stupid to understand?"

By this time, my weight and my rifle were causing me extreme trouble. I straightened my knees and released my hands simultaneously, my rifle hitting the floor of their room. Sgt. Swan turned me and shoved me against the wall. Physical hazing was against the rules, but it was a rule that was bent more often than anyone in charge ever knew.

"YOU JUST LET YOUR RIFLE HIT THE GROUND, MAGGOT. THAT RIFLE HAS BEEN IN THE TRENCHES. IT HAS SAVED LIVES AND MADE THIS COUNTRY FREE SO THAT WHORES LIKE YOUR MOMMA HAVE THE RIGHT TO SPAWN IDIOTS LIKE YOU!!! YOUR RIFLE SHOULD NEVER TOUCH THE GROUND WITHOUT BEING PLACED THERE WITH CARE AND COURTESY."

There was more. More than I can remember, and more than I should write. I am certain you can get the gist. Cpl. Smith was right beside him, yelling at me with equal gusto. The blond moron in the chair was still in the chair at the desk, but laughing and overall enjoying the show.

"If your rifle is going to lay on the ground, you are, too. Now get on down," Sgt. Swan ordered.

I dropped to the front leaning rest, preparing to do the push-ups that I was getting quite good at by now.

"I want you on toes and elbows maggot. If your rifle has to be uncomfortably lying on the floor, you do too."

Well, if you have ever held a plank with your body resting only on your toes and elbows, your chin in your palms, you can imagine my state. The relief of releasing the door was well behind me now. Within minutes I was flat on the floor, having reached muscle failure again, only this time with different muscles.

"What are we going to do with this one?" Sgt. Swan asked Cpl. Smith.

"I'd say there isn't much that can be done with him. If only his mom had been on the pill, it would have saved all of us a bunch of trouble."

"Get up, maggot," Sgt. Swan said, pushing the toe of his tennis shoe hard into my ribs. "I guess you better sit down."

Relief washed through me. They had seen the hazing was going too far and enough was enough. I was finally going to get to sit down.

"You'll get to spend some time in the green chair," Sgt. Swan said.

This was my first time using the green chair, which is a wall-sit with your back against the wall, legs bent so that your thighs are parallel to the ground and your shins are perpendicular.

"Arms straight out," Sgt. Swan said, and he placed my rifle in my hands at the end of my extended arms.

Again, I was at muscle failure soon as my bottom slid to the floor.

The positions were repeated, again and again, from the locker door to on the floor to the green chair. No relief in any of them other than the briefest seconds when I got to leave one position for the next. The music continued its assault as well, I think. To be honest, I can't remember if the music was still playing throughout or had been silenced.

"You better get out of here before we decide to get rid of you for good, maggot," Sgt. Swan said. He said it quiet. Full of threat. "Get your stinking body off the floor and get out of my sight before I decide to do something that only your mother will regret. Cpl. Smith, I think A Team will be one lighter tomorrow."

I left their room, a quivering mess. I made it up to my room to find out that it had been less than an hour, but less than an hour that was burned deep into my memory.

My next recollection of that night was at about 2300 hours, 11:00 pm for civilians. Taps sounded at 2200 hours, and it was lights out. I hadn't been able to sleep. The beam from the TAC officer's flashlight had passed over my still form, faking sleep for bed check. I had been waiting for that to happen so that I would have the largest amount of time while he continued his rounds of The Box.

I snuck out of my room and down the stoop to where the pay phones hung in the corner of The Box where Delta Troop met Headquarters Troop. Getting caught would have been a disaster, so I was being quiet and moving in the shadows.

This was a time in America when you still had to put a quarter in the pay phone, and you would raise a real live American human operator when you did it.

"AT&T, may I help you?" the angelic voice of an obviously beautiful and benevolent night operator said.

I sat there on the concrete under the pay phone. Holding it to my face, pressing hard. Tears on my cheek. If I just made my folks know how hard it was here. How tough. How mean the old cadets were. The chance that I wasn't going to make it out of here alive. I just needed them to come to get me. Right now. Tonight. The ranch was only three hours from Roswell.

"AT&T, may I help you?" the sweet voice said again.

My voice caught in my throat, but there weren't any words.

"Is someone there?" the voice said.

I stood up and hung up the phone. Wiping the tears off my face I took the returned quarter from the slot and decided right then and there that they would not break me or make me leave. Heck with them. They were going to have to kill me to get rid of me, not that that wasn't a real concern.

Walking back to my room, my resolve had hardened. I walked

back as if I didn't care about being caught. Head held high and iron being woven into my will. The risk of a TAC officer catching me meant nothing to me as I walked down the center of the stoop instead of against the rail where RATs were supposed to be. I was no longer that little kid who had gone down to Sgt. Swan's room looking for a friend. Now I was Cadet-Recruit Gregory, C. L., 830087, Delta Troop, A Team, 2nd Squad, 2nd Platoon. I was stronger at that moment than I had ever been.

0520 rolled around and I got out of bed with determination. I made my bed, got dressed, and was on Sgt. Swan's threshold when reveille sounded at 0600, his spit-shined shoes in my right hand, his polished brass placed in his clean shirt in the other. Mr. Coles, a RAT from Sgt. Swan's roommate's squad, was right beside me with his squad leader's shoes and shirt as well.

"Cadet-Recruit Gregory, C. L., requesting permission to enter," I said.

"You're still here?" Sgt. Swan said from his bunk. "Sgt. Deering owes me a Coke. He was my buddy from India Troop that was at our little party yesterday. He said you'd be gone this morning, but I bet a Coke you wouldn't. Good job, maggot. Maybe you're just too dumb to leave?"

I placed his shirt on the chair, his shoes on the floor, and handed him his Copenhagen so that he could start his morning right with a dip of snuff before he got out of bed.

"Still here, Sergeant, and no plan to leave."

"Good for you. Maybe we'll have to see about that plan of yours."

The promise of more hazing was evident, but mentally, it didn't matter to me. I had gone from the scared sad kid to the attitude of "bring it on".

He placed a dip in his cheek, and Mr. Coles and I waited for our respective squad leaders to get up so we could clean the room and make the beds.

At least I had won my squad leader a Coke. Things were looking up.

32

OLD CADET CORNERS AND GRASS

The concrete troop walks went from the bottom stoop of the barracks to the middle of The Box. The walk was large enough to form up over a hundred men, probably measuring twenty feet wide and one hundred feet long. These all led to a central avenue through The Box that ran the whole length of The Box, and I would guess from memory they were about fifty feet wide and three hundred and fifty feet long. Those distances are just a guess from memory and may be completely off. At the corners, there was a triangular piece of concrete that made the ninety-degree meeting between the walks into an angled, or softer, corner. These softer corners were only for old cadets to walk on, never for any RAT at any time. Even if you were on a mission from God (being sent to get something for an old cadet), no RAT was allowed to even think of cutting a corner and touching the old-cadet corner. The old cadets had a similar affinity to their grass. The grass part I understand, since it is such a rare and difficult thing to find and maintain in New Mexico, but the old-cadet corners were concrete and maintained their durable nature despite how much they were used.

"Cadet Gregory," I heard a voice call. I stopped instantly, going from the light jog required of every RAT movement in The Box to a dead standstill.

I didn't say anything, because, at this point, I knew the difference between an officer and a noncommissioned officer and did not want to inspire the torture that came from either one when they were misidentified. Without having seen the caller, it could have been anyone.

"About face," the voice said again.

I executed the move to see Sgt. Gain and another NCO that I did not recognize walking toward me.

"Our pizza is arriving in thirty and I need a RAT to go get it," Sgt. Gain said.

Domino's Pizza delivered to NMMI regularly. With the Bates Hall Den of Culinary Atrocities living up to its name, the need for additional calories that were edible was real. RATs did not often have the luxury of getting a pizza for themselves, but at least we got to fetch them for old cadets.

"Yes, Sergeant," I said. "I'm headed to the bookstore for something I need for class, Sergeant. Can you please find another RAT?"

"You've got some mouthy RATs in Delta Troop," the unidentified sergeant said to Sgt. Gain. "If one of mine from Foxtrot talked back like that, I'd have to make an example of him."

Well, this was not looking good for me. Sgt. Gain was in 1st Platoon, and I had yet to have any dealings with him. The other one, I was about to find out, was named Sgt. Owen. He had that look of mean that RATs came to recognize from experience and exposure. Probably the same thing prisoners learn to recognize in the expressions of the guards.

"I reckon we'll get a little RAT race on," Sgt. Owen said. "Maggot, run to Foxtrot and bring me the first RAT you see. Double time."

My trip to the bookstore was now on hold for certain and indefinitely. I ran across The Box to Foxtrot and found a college new cadet on the first stoop. "Hey man, Sgt. Owen wants you."

"Oh no," the new cadet said. "Not him. That guy is demented."

"Sorry, he sent me to get you," I said.

"Did he ask for me by name?" the new cadet asked.

"What's your name?"

"Davis," he answered.

"No, just any new cadet," I said.

"I ain't going," Davis said.

"OK," I said. "What do you want me to tell him for you? He's in a hurry, so you can come or give me someone else. They're watching us right now," I pointed over my shoulder.

"Dang it, dang it, dang it," Davis said. "OK. Well, we're in for it now."

We ran back to Delta Troop and reported to Sgt. Gain and Sgt. Owen, standing at strict attention waiting for their command.

"Let me teach you how to handle a RAT that won't do what he's told," Sgt. Owen said to Sgt. Gain.

"OK, maggots," Sgt. Owen said. "You're going to have a RAT race and the winner gets to get our pizza. You do not want to be the loser. ABOUT FACE."

I have never been the fastest or most athletic, and by the looks of Davis, that was about to be well-documented again. Rodeo was my chosen sport since the horse did the running for you.

"You maggots are going to run down and back across the grass, responding to orders as I call them," Sgt. Owen said. "GO."

Davis was off, knowing what to expect. I pursued, already behind. "DROP," came the order.

Davis hit the grass and started doing push-ups. From ten feet behind him, I did the same. "CRAWL," so we were now low crawling.

Both of us were in PT uniforms, so the low crawl was not fun or easy on the naked knees. "RUN." So, we did.

"DROP," followed by "ROLL," accomplished by rolling toward the finish line, and another "CRAWL," was thrown in. By the time we made it back to the first stoop of Delta Troop, Cadet Davis was at least thirty feet ahead of me.

As I came to the end, I had added insult to injury by accidentally getting part of one foot on the concrete of an old-cadet corner. A bad situation had just become worse.

"Looks like the best man won," Sgt. Owen said to Sgt. Gain. "I'll teach you how to make these RATs stronger. Go get the pizza, maggot," he said to Davis, handing him some money.

Relief washed over Davis's face as he took the money and turned his dirty uniform and bleeding knees away from me in a perfect left face and started running to Sally Port where the pizza would be arriving.

"You just stepped on my corner, you %$&*ing idiot," Sgt. Owen said. He was close to my face, speaking in a low tone that was more frightening than if he had been yelling. "You know that hazing is not allowed?" he asked.

"Yes, Sergeant," I said.

With a short punch, he hit me in the stomach, doubling me over. "That's not hazing, maggot," he said, leaning down to stay close to my ear. "That is discipline and negative reinforcement. You just walked on my grass and hit the edge of my old cadet corner. Have you suffered brain damage?"

I was gasping for air, so I didn't answer. Our race had been on the grass in the first place so I shouldn't be in trouble for that. At least that is what I thought.

"Attention," he ordered.

I stood back up as best I could, still hurt from the punch. "About face." I complied.

"Front-leaning-rest position. Move."

I went down into position, my hands on the walk and my face over the grass. Incidentally, the implied rest in the name of this position is part of the reason why so many cadets end up with trust issues.

"Down," he ordered. "Kiss the grass and apologize to it for letting your maggot foot touch it."

I was already close to muscle failure from the race I had just lost, but I tried to go down without collapsing.

"I don't hear you, maggot," Sgt. Owen said.

Out of breath, "I'm sorry for stepping on you, you beautiful grass," I managed.

He made me repeat this for a while, demanding creativity as I apologized to the grass.

"Crab walk over here to the old cadet corner," he ordered.

I moved over as he commanded, to kiss and apologize to the concrete. I complied the best I could, trying to talk, trying to kiss, and trying not to hit the ground with my exhausted arms when I heard, "Back off."

"What's it to you, Private?" Sgt. Owen said back to the voice.

"Leave him alone," the voice said.

"You going to make me?" Sgt. Owen asked.

With that, I heard a scuffle and a body being slammed against the iron rails that lead up the stairs.

"Get up, Chris," the voice said to me. "I told you to back off. I've been watching and enough is enough."

It was rare to hear your first name. Even amongst other RATs, your last name was on your nametag and kind of became your first name while at NMMI. At least I had two first names.

I stood up to see Pvt. Daem pinning a very frightened-looking Sgt. Owen against the rails. Pvt. Daem was a stout and short bull rider on the rodeo team and had been at NMMI for a long time. He had no regard for the military side of things, as evidenced by his rank after four years in the corps. He was as tough as anyone I have ever known and was not someone that you would want to cross. His thick forearm was across Sgt. Owen's chest, even though Sgt. Owen had an easy eight to ten inches on him.

"I am going to make you," Pvt. Daem said. "Now you apologize to Cadet Gregory for the misuse of your rank."

"&%^$ you," Sgt. Owen said.

A mistake. Pvt. Daem sent a short punch into Sgt. Owen's solar plexus, harder than the one that I had received. He couldn't double over because Pvt. Daem was holding him up, and he couldn't catch his breath either. Sgt. Owen's eyes started to water, and he was gasping for breath.

"Are you ready to apologize?" Pvt. Daem asked.

I was standing there watching this all unfold in disbelief. Sgt.

Gain was in our troop and knew that, regardless of rank, Pvt. Daem was not to be interfered with. I don't think even the Regimental Commander would have bothered Pvt. Daem.

"Last chance," Pvt. Daem said.

"I'm sorry," Sgt. Owen managed to gasp.

"That's so nice of you," Pvt. Daem said to Sgt. Owen, letting him bend over to hide his crying and relieve his gut. "I can see that you are smarter than you look."

Pvt. Daem leaned over to get close to the back of Sgt. Owen's head. "If you so much as look at Cadet Gregory again, you will regret it more than anything you've ever done. Do you completely understand me?"

Sgt. Owen nodded his response. Turning to Sgt. Gain, Pvt. Daem said, "Goes for you, too, Jeff."

Sgt. Gain was just as agreeable as Sgt. Owen, his head bobbing up and down like a bobblehead doll. Cadet Davis had returned with their pizza, and they headed across the Box to get away from Pvt. Daem, Sgt. Owen wiping his eyes and walking with a stoop.

The rank system allowed for sanctioned bullying, which is not a new or uncommon thing. Thankfully, there were those among us that could see when a line is crossed and keep the sadistic natures of some in check. Rank played no part in this instance once Pvt. Daem stepped in.

Pvt. Daem was the kind of man who knew right from wrong and was willing to do what it took to make sure that right was done. He saved me on more than one occasion, the next time dragging me off a bull that I was hung up on.

33

RODEO PRACTICE

"Your bull's loaded," Nat Pat said.

I climbed onto his back and settled in with my feet on the slats of the chute. Rock pulled up on my bull rope, and I rubbed it up and down to warm the embedded rosin and make it stickier. Pinching the rope to the handle, I wrapped the tail around the back of my hand, squeezed tight, leaned over my hand, and nodded my head for Nat Pat to release the gate and let us out into the arena.

To finally be at the rodeo grounds was great. The whole military side of military school had been tough, way tougher than expected. At least at the arena, we were allowed to wear "civies" (civilian clothes) and kind of feel like we were part of the world we had come from.

The NMMI rodeo team was one of the few places where a high school rodeo competitor was part of organized practice. As such, we were dominant in the high school rodeo association that held rodeos across the state. I had wanted to be a bull rider in the worst way, but I had found through the increased exposure I was getting at NMMI that I was not very good at it. At home, I had only been on bulls at rodeos since we didn't have an arena or any bucking chutes. Trying to learn by only competing is not a great way to get good at anything, which was my main reason for wanting to go to NMMI in the first place.

NMMI had a bucking machine that we could use a lot, as well as a pen of bulls that we rode at least twice a week during good weather. I was not a natural bull rider. Some guys were, Pvt. Daem being a great example. They just looked comfortable and correct. Even when I did cover one, I looked more like I had gotten my hand accidentally stuck to a bull instead of the controlled and precise movements that some of the other guys were able to do.

The bull I was on here was a big brindle bull with quite the horn spread. We called him Brindle, the red bull was called Red, and the black bull was called Casper the Friendly Ghost. Brindle was generally nice when he bucked you off, which I felt was a good feature for a practice bull, especially ones with horns like he had. There were enough that would come looking for you when you hit the ground, like Casper the Friendly Ghost. Casper was a headhunter. Most bull riders are quick to find their feet and get out of the arena when the ride is over.

Pvt. Daem was in the arena to act as a rodeo clown. I nodded my head, Nat Pat untwisted the rope that held the gate, and Brindle and I quit the earth. As the gate swung away to my left, Brindle hit the gate with his horns and sent it smashing back on its hinges to crash into the chute in front of us, sounding like a car wreck. Nat Pat pinned it against the chute it had hit so it wouldn't be in the way of Brindle's efforts.

He spun to the left and I felt my body going down into what is known as "the well." The well is the inside of a spinning bull, where the cowboy ends up if he counteracts inertia too much with his free arm. Inertia is trying to throw you to the outside of the spin and there is a balance that you are trying to find that allows you to stay on top of the bull. I felt a horn catch my left arm and pull me down, and I was off the left side of this monster, away from my hand. "Away from your hand" is a term in bull riding that refers to coming off the opposite side of the hand that you ride with. In my case, I rode with my right hand.

My hand remained in the rope as Brindle came out of the spin and started across the arena. I felt a hoof step on my right foot and someone grabbing my left arm and shoulder. The next thing I remember is being at the infirmary.

"Daem saved your bacon," Rock told me as we stood in the waiting area of the infirmary.

"How did we get here?" I asked him.

"You don't remember?" he asked.

"No, what happened?"

"You hung up on Brindle and Daem pulled you off."

"How did we get here?" I asked again.

"You were a little loopy when you got back behind the chutes, but you didn't pass out. After watching you for a while I thought I had better have the doc look at you. You walked here with me, whether you remember it or not."

"Did I ride him?" I asked.

"Maybe for all of three seconds," he laughed. "Your boot is ruined, and your arm may be a bit sore by the looks of it."

I was starting to take inventory of my injuries. My right shoulder was the worst, and that injury has followed me even until today. My left elbow was hurt, and there was a big bruise on my left forearm where Brindle must have caught me with his horn. My head was sore, but I didn't feel any bumps or have any blood that I could find. My lower back had suffered a tweak, and I have always blamed any back injuries on that damage, probably unjustly. The back of my right heel was also sore, and I found out later that the heel of my boot was missing where he stepped on it and broke my spur strap.

Rock and I were both in PT uniforms, so I had gotten dressed and made it back to the main post somehow, even though I didn't have any recollection of it. The nurse checked me in for the night and I hobbled around for a week or two as the aches and pains slowly subsided.

About three weeks later, I was ready to get on another bull, or so I thought. I had been on the bucking machine and had been roping again, but this was going to be my first bull since getting hung up.

They loaded Red for me. He was the easiest one in the pen to ride. I climbed down on him. This little bull had been a confidence booster, and we all knew, without saying it, that this was why I was getting on him. Rock pulled my rope as I warmed up the rosin, Nat Pat was on the gate and Daem was in the arena. *Déjà vu* from the last bull I had been on, other than the bull itself.

I nodded, the little bull turned to the left with a small kick and hop, and I was on the ground. I don't remember telling my hand to let go, but the fact was, it had let go. Everyone was nice about it, and I tried again a few more times over the next couple of weeks or so, without success. In fact, I can't remember ever covering another bull after being hung up on Brindle.

It is with shame that I must admit that I was not able to mentally overcome the wreck I had been in. This is one of those things in my past that still haunts me. Knowing what I know now, I have developed my mental game, and I would have been able to get over it. That knowledge has also aided and allowed me to coach and help my kids and students over the years, but the ugly fact remains that I personally was not tough enough to become the bull rider that I wanted to be when I was fifteen years old.

When my son Cody was a young bull rider, he was tougher, stronger, and more natural at it than I ever was. I remember a conversation with my dad, and he said, "Well, it might be better if those kids weren't so good at rodeo." He was completely right about that. Old age has taught me there are things to enjoy and things to excel at. It is such a great thing when those can be one and the same, like farriery has become for me. It's even better if it pays well. Rodeo does not pay well for most of its devotees.

For the rest of my rodeo career, I focused on steer wrestling as my main passion. I roped, as well, but what truly got me motivated was the feeling of jumping on the head of a running steer and throwing it to the ground. This ended up earning me a college scholarship for a couple of years, and I even got to enjoy the sport into adulthood for a time. Not becoming a great bull rider is one of those unanswered prayers that old men look back on and see as God's blessing and guidance, even when they didn't understand it at the time.

34

EPAULETS

Sometime before Thanksgiving, the new cadets that hadn't quit were initiated into the Corps of Cadets. It was a night of tradition and ceremony, and we were all given epaulets for our uniforms. To this point, our uniforms had remained conspicuously naked at the shoulders where the buttoned strap that secured the epaulet was sitting there, lonely.

Our epaulets were nothing to look at. They were simple black pieces of cloth shaped like the cardboard from a roll of toilet paper, just smashed and black, and the epaulet holder passed through it. Once a cadet had rank, the epaulet told of his authority. Even just one yellow boomerang-shaped chevron of a private was enough to make life endurable.

"Never thought you'd still be here, maggot," Sgt. Swan said as he buttoned my epaulet to my uniform. "You may be too dumb to leave, as I've said before."

I sure didn't care to have Sgt. Swan touch me in any way, but I just stood at attention and didn't say anything.

"Welcome to the corps," he said. "Give me twenty."

I started doing push-ups as were dozens of other RATs in every troop.

"Fall in," 1st Sgt. Wengert yelled.

We ran to the troop walk and formed up in our designated spots. "Welcome to the corps," he said. "Half-right, face," came the command.

The old cadets stepped out of formation and just the RATs were left. We all knew the command of half-right, face, was given just before being dropped as a platoon or company. "Give me some push-ups for your initiation," 1st Sgt. Wengert yelled. "ONE – TWO – THREE – FOUR – FIVE...." He continued until he was bored with us. I was up to well over fifty push-ups so far and we hadn't even started toward SRC.

Once the whole regiment was formed up and all the troops were ready to head to supper, the Regimental Commander said a few words to welcome the RATs to the corps and explain how this meant something. The date was far enough into the school year that the weakest had been culled and only the deserving remained.

When supper was over and NSH (night study hall) was set to start, old cadets were on the prowl to drop new cadets and welcome them with a zoom broom. The zoom broom was a normal broom with the bristles all taped together with duct tape and the handle cut off. It resembled a paddle more than a broom.

"Assume the position," Sgt. Johnson said.

I had hit a wall when he walked in and was standing at strict attention. "What is the position?" I asked.

"Don't forget your place, maggot. If you have a question, request permission to ask it," he said. "And never address me again without using my rank. You think getting initiated into the corps gives you license to talk to an old cadet like a friend?"

Putting my fist forward, I said, "Excuse me, Sergeant, Cadet-Recruit, Gregory, C. L., requests permission to ask a question."

"Go ahead."

"What is the position?" I asked.

"Grab both sides of the heater and don't jump," Sgt. Johnson said, grabbing the heater by the window in my room to demonstrate.

I repeated and he smacked me across the buttocks with the zoom broom. My roommate followed suit and we endured the first of many to come that night. Old cadets were wandering around the barracks like trick-or-treaters on Halloween, administering a zoom broom welcome at every stop.

While the zoom broom was bad, it paled compared to the officer's saber. All the platoon leaders, troop commanders, and senior officers carried a thin saber as part of their uniform. Two new cadets were bent over side by side and made to grab their ankles. The officer would hit the closest one with the middle of the saber blade and the tip would whip around and strike the second cadet with speed and angle. The RATs would then change places and the officer would repeat the action so that both could endure each position. This would leave a triangular bruise that made the zoom broom feel like a lover's caress. It was a brutal evening, and I began to understand how my NMMI alumni Ag teacher, Mr. Scott, had become so good at swinging his paddle named Norm. I am ashamed to say that I dealt out my fair share of the zoom broom when I was an old cadet.

35

PELVIS BUCKS ME OFF

Inspections also meant a parade. Not in the Rose Parade sort of way, but in the military way. The parade grounds were a large lawn between The Box and the rodeo grounds where 1000 cadets could form up. Every week, this lawn was used for "drill" where we would all practice marching and moving about as a platoon. All the cadets were marched around learning how to stay in step and how to do flanking movements, rifle drills, and basic movements. When it was time for a parade, the whole school would be formed up in their troops and there would be a pass-and-revue.

NMMI also had a cavalry unit that included about 40 horse-interested cadets. They had half of the stall barn where the rodeo grounds were, and they got to practice cavalry movements so that they could be in the parade as well.

As a rodeo cowboy, I was able to be on the Mounted Color Guard (MCG). We formed up in the middle of the parade ground with six horses. The riders on each end carried pistols, and the four in the middle carried flags. There were several advantages to being on the MCG. First, we didn't have to be in the inspection, so our rooms ended up being locked and full of everyone's contraband or extra stuff that they wanted out of their room for inspection. The other benefit was that we didn't have to stand in the parade and got to go out to the arena and get our horses ready for the parade while everyone else was going through a not-so-fun inspection.

As in all things military, the saddles had to match, so we weren't always riding our own western saddles. We ended up riding the old McClellan cavalry saddles. They were not as comfortable as our roping saddles nor as easy to stay sitting in if your horse didn't agree with the whole situation. When I became MCG Commander, my senior year, I would get the approval for western saddles for at least half of the parades. Evident in the cover photo of this book.

I had a horse named Pelvis, so named for bucking a gal off and breaking her pelvis. He was big and strong and not necessarily that well-trained. I roped off of him, but Pelvis was not always the most agreeable partner. Most of the time, he would buck a little the first time he was ridden each day like he had forgotten what had happened every day before. He was fantastic for a game we played called pull-off, and he would have been an exceptional pick-up horse. You just had to get past the first minute.

"You're carrying the NMMI flag," Lenny told me. He was the MCG commander my RAT year.

"Sounds good," I said. "I haven't been on Pelvis in a real parade yet, just practice."

"No time like the present," Lenny said. "You've got him warmed up already, so you should be good."

The horses had already been ridden and worked that morning to get them ready for the parade. We groomed and saddled our horses with the McClellans, got all dressed up in our uniforms, and mounted up to take our place in the middle of the parade grounds. I was sitting taller than anyone else on big ol' Pelvis.

Whatever general was giving speeches gave his speech. We all stayed where we were assigned, and it was finally time to start the parade during which the whole regiment would march past the bleachers.

"PISS IN YOUR SHOE," the regimental commander shouted. His order rang across the parade grounds and was a joke that he could later explain to a real Army officer in charge that he said, "Pass-and-revue." We all knew the truth though.

The battalion commanders passed the command back to their troops. Starting with Alpha Troop, the troop commander ordered, "March," and the troops started off. We stayed where we were, and the Pixies passed behind us. As a fun exercise, the drummers exploded with extra gusto when they were behind the horses to see if they could get a reaction. Pelvis was happy to oblige them on that day.

Suddenly, Pelvis and I were headed across the parade grounds as he bucked and leaped to escape the horse-eating drums. My fellow MCG cowboys were engaged in keeping their mounts under control as Pelvis and I broke formation and headed toward the mass of marching cadets. I got him turned, I don't know how, and we headed away from everyone as I lost the NMMI flag, lost my stirrups, and eventually lost my seat. Hitting the ground hard, I stood up and recovered my flag as the pistol guards recovered my horse.

Lenny led Pelvis back to me. "Get on," he said. "This isn't any time to be running around and messing about."

"Yes sir," I said.

He smiled, kind of, at me, even though he was a sergeant and I'd been there long enough to know the difference. "You're gonna owe me for this one Gregory. Not the time to be playing cowboy."

I knew that. Pelvis should have known it too. He had caused me some real trouble, but it wasn't my fault. Nothing I could do about that now.

"What's your name," a lady hollered at me as I was just about to ride back to the MCG formation.

"Why?" I asked as she continued toward me.

"I'm a reporter for the *Roswell Daily Record*," she said.

"John Smith," I answered without really thinking about it.

The NMMI has an honor code, "A cadet will not lie, cheat or steal, nor tolerate those who do." This was the code we were to live by, but I had just broken that lie part of it.

The next day there was a photo in the paper of me climbing back on my horse, his huge body lifting the McClellan saddle stirrup farther from the ground than it should be as my left foot reaches for it.

The caption read, "Horse ruins the day for Cadet Smith."

It was an honor violation, but not one that was malicious or intentional. I just responded to the embarrassment by hiding my identity. Luckily, no one called me out on that one, and I was able to stay at NMMI for another two and a half years.

36

STREAKING

A tradition had formed in the crazy times of the 1960s that involved streaking. Every year in April after the Annual Formal Inspection (AFI), the corps would go wild. About 500 cadets would hit the streets of Roswell, running through the town without the benefit of cloth between ourselves and the elements. I went streaking three times in three years at NMMI, but they were cracking down hard on us in my junior and senior years, so both of those were a letdown compared to the first year.

"What are you wearing?" Rock asked me. "I'm just wearing chaps and a hat."

"I'm going to wear my cowboy hat and a half shirt," I said.

"Colby is going completely naked," Rock said.

"I heard a bunch of the cowboys are," I said. "Guess it don't really matter. If you're not wearing anything to cover up the middle, don't matter if the rest is covered up or not."

Night study hall ended, and the corps left their rooms as one. Several other cowboys and I headed down the main street that led in front of the school. A van with a couple of girls came along and picked us up. Many of the townies were more than happy to help facilitate the fun, so they would take the cadets to businesses that were to be given a show. My group went through Denny's, a hotel, the movie theatre, and McDonald's. When we were dropped off close to the school, we were hosed down by the fire department that had been called out to stem the fun. The local radio station was announcing the streak and talking to observers that called it in. Some roads were lined with people that were cheering us on as if there was a parade in town.

Girls had been allowed to attend NMMI since sometime in the early 1970s or so. They were all in Kilo Troop, but there weren't that many Kilo Troop members in my day, maybe fifty or sixty, at the most. Kilo Troop was not housed in The Box with the rest of the troops. Their barracks area was called The Slab and contained Kilo, Lima, and Mike troops, with Kilo being all female. They didn't join in the streaking fun as runners, but we did finish up with a dance in front of Kilo Troop as the streak ended.

"I'm sore as can be," Ronnie said the next day in English class.

"No kidding," I said. "Streaking may be the greatest workout ever invented. I think all of us ran harder and faster than we ever did in PT."

"Colby was arrested, buck naked, hands cuffed behind his back when they caught him climbing off the roof of the public library," Ronnie said.

"I'll bet that was an uncomfortable way to be processed at the police station," I laughed.

"Colby was bound to like it more than not," Ronnie said. "He's a California showboat if there ever was one."

Colby later confirmed that it was a lot of fun. He claimed that the cops left him handcuffed until after he had been booked but let his hands be as free as the rest of him before giving him a long shirt and putting him in the cell. He didn't get in severe trouble but did end up with some serious tour duty out of it.

37

MEMORY BUILDING EXERCISES

Information was thrust upon new cadets from every direction at such speed that there was no possible way to take it all in. Even the smartest and most gifted cadets were constantly challenged. Learning how to shine shoes, place brass, clean rooms for DRI (Daily Room Inspection), make the bed properly, stand, march, move correctly, eat at a brace, request permission correctly for everything, know where to walk, know where to run, understand all corps vocabulary and military customs, etc., is just a short list of things that were part of corps life. Add to that an academic curriculum that was well beyond what any of us had experienced in public school, and the learning never stopped.

At a squad level, there were sayings that we had to remember and repeat verbatim when asked certain questions. It was nice that they had given us something to do since we needed a way to fill all that spare time. Geesh. There was also a great way to help that young cadet remember these little gems.

"Cadet Gregory, what time is it?" Sgt. Swan asked. It was BRC on a brisk October morning, and I was ready to march to breakfast with the troop.

"I haven't learned that one yet, Sergeant," I answered.

"How can you not remember that one?" Are you touched in the head? Did your mom breastfeed you or were you raised with beer in your baby bottle? I don't think I've ever seen a dumber cadet. I gave you that paper last night and all you had to do was remember a few simple words. Let's get on down and I'll help you learn this."

With that, I was in the front-leaning rest position and was doing a push-up with each word as he said it slowly.

"Let's start pumping," Sgt. Swan said. "Sir, I am deeply embarrassed and greatly humiliated, but due to... You are not keeping up, maggot. Either learn this thing or start doing your push-ups faster."

I was going as fast as I could, and push-ups had become so much easier in the last few months, but there was no way I was going to keep up with his tempo. Slow as it was, it was fast enough to keep me from keeping up. In every squad, similar events were being played out as we were learning just how important being able to remember these nonsensical sayings were going to be.

Immediate punishment for slow learning is a fantastic way to create some long-term memory of any topic. That is a trick that I have successfully used with farriery students for my whole career. Push-ups lead to compliance.

The amazing thing is that I still remember these things today. Almost forty years later, I can say them with the same speed and accuracy as my fifteen-year-old self learned them. I'll share a couple of my favorites here.

Question: "Cadet-Recruit ____, what time is it?"

Answer: "Sir, I am deeply embarrassed and greatly humiliated, but due to the hidden components of my chronometer over which I have no control, and the greater sidereal movement by which time is commonly reckoned, I can with no great accuracy state the exact time. I can, however, without fear of being too far from the truth, state that the exact time is zero-six-one-five hundred hours, sir."

Question: "Cadet-Recruit ____, why are fire engines red?"

Answer: "Sir, two plus two is four, three times four is twelve, there are twelve inches in a ruler, Queen Elizabeth was a ruler, she ruled England, England had ships, ships sail on the sea, there are fish in the sea, fish have fins, the Finns fought the Russians, Russians are red, and fire engines are always rushing, sir."

38

RAT YEAR COMES TO A CLOSE

After the first nine months as a high school RAT, if you had survived, you became a private or private first class. College RATs only had to suffer being a RAT for one semester, so most of them had been lording it over the high-school RATs since Christmas. It was a glorious promotion when you got to go from a life of being bossed around by everyone to becoming one of the bosses. Now you could walk to class. You didn't have to hold the door open for every old cadet. You didn't have to request permission to eat, walk, run, stop, move, poop, etc. Table manners went out the window in an appalling manner, as if we had all been waiting to see just how badly we could act in front of a tray of food. It was wonderful.

There is a principle of leadership and control that should be recognized here, one that I have seen many times and have used many times as well. The principle is this: if people are denied all their rights, then any small right they are granted seems like the biggest gift and garners appreciation well beyond its value.

A good example is when there was a blood drive at NMMI, which there were every fifty-six days to the hour. Any cadet who gave blood was granted a "free rev." Basically, that meant they did not have to get up with reveille and got an extra hour of sleep. A massive percentage of troops would gladly spend a couple of hours of their day and give a pint of blood for the promise of sixty minutes of extra sleep the next morning. And then, we would all go about saying how great it was to get a free rev. What a generous and benevolent place we were living to have traded an hour of snoozing for spending just two hours standing in line, answering a bunch of questions, and being stabbed by a pipe posing as a needle to take a pint of blood.

Becoming a private was like that. The fact that you didn't have to run everywhere was so fantastic that you felt as if you had been released from death row or something. Used properly, this principle can teach a lot and build a lot of character in a youngster, a soldier, a student, or anyone over whom you may have influence.

The only problem was that the school year was only days from being finished, so none of us had any opportunity to wield this newfound power until after summer break. There were no RATs to try it out on. But we all went home knowing that life would be different when we returned in the fall.

Once you became an old cadet, life was so much better at NMMI on a massive scale. When you wanted something from the PX, or anywhere else for that matter, you sent a RAT. RATs that were sent on any sort of mission were able to run through some of the places that were only allowed for old cadets. If stopped, they would say, "I'm on a mission from God," and were then permitted to pass. Being an old cadet was like having dozens of personal assistants.

Every morning, your room would be cleaned and your uniform laid out by RATs. In the winter, you could have a RAT sitting on the toilet so that the seat would be warm for you (wearing his underwear of course, as if anyone would let a RAT's naked bottom on a toilet they were about to use). Looking back, most of us were not applying or using great leadership as I recognize it today. We were being cruel like teenage boys can be, but we had been taught these techniques for nine months as we served our time as a RAT. To the credit of the old cadets who had been in charge of us, they were also only doing as they had been taught in the same hard way that we had been taught.

It was a different time with different standards, and I feel that it made for stronger men. We survived, and so did those who came under our control. When I later enlisted and went to the real US Army Basic Training, it was so much easier than life had been as a RAT at NMMI. Basic training seemed just like camping and fun with friends. There were many soldiers with me in basic who saw it as the toughest thing they had been through to that point in their lives, but to any NMMI alumni I talked with, basic training was almost a joke.

Throughout my adult life, as I have told these stories to civilian friends, I have often had a person comment, "They would never have done that to me," or "I wouldn't take that. I'd show them who was boss." The truth of the matter is that some of the biggest, strongest, and toughest people have gone through much of what I am writing about. They came through it tougher yet because of the experience. Those who did not comply were eliminated, no matter how tough they thought they were. Things like this were easier to endure because your sweat and blood were mingled with that of the man suffering next to you. You were not the only one being humiliated, and you had your buddies to commiserate with later. It is hard to explain and hard to imagine unless you have been through something similar. I think it is a wonderful part of being a human and being part of something bigger than you and in a place that strengthens every part of you.

39

LITTLE MIKE, THE SOLDIER

My brother had been a year behind me in school, and when my second year rolled around, Little Mike joined me at NMMI. I got my growth early in life, so by the time I was fourteen, I was already six foot one. Eating five cheeseburgers at a sitting had helped that happen. Little Mike didn't get his growth until he was in college, so we looked years apart instead of only sixteen months. The top of his head was not quite to my shoulder.

It was such a great thing to have both of us there, and Little Mike proved to be one of the most natural soldiers ever. I don't remember coaching him or giving him a lot of advice, but I must have, because he thrived at NMMI. Or, maybe he is just that talented. That might be it as well.

Little Mike is also gifted with a way of telling stories that accentuate the funny, and if that means the facts are exaggerated, the end is worth the means. It would be such a treasure if he were to write of his time at NMMI. It would certainly eclipse my tales.

By my senior year, I was only a staff sergeant they promoted to sergeant first class graduation week only because they knew I was not going to be there for college. By my brother's senior year, he was his troop's first sergeant, which is a very rare thing for a high school student to achieve. Only the best of the best became a senior NCO as a high school student. Most of those spots are filled by the junior-college cadets who were going into the regular Army. I was, and still am, very proud of him.

Having him there meant we got to team rope together. Mike is an outstanding header (the cowboy who ropes the horns and turns the steer for the heeler to catch the legs). I was the heeler, but I was way more interested in throwing them down than I was in roping them. I was the weak link on our team, but we did have a lot of fun in the arena together. Mike went on in his adulthood to ride cutting horses and has achieved results far beyond what most could ever dream of. He is a very natural horseman.

Even though we were not in the same troop, we were both on the rodeo team and the Mounted Color Guard, so we got to see quite a bit of each other. Our folks went to every high school rodeo. We were allowed a lot more freedom on rodeo weekends than any of the other cadets would have believed, so each rodeo was a glimmering little oasis of hope and memory of the world we had left behind. Mom would have coolers of food, and all the rodeo team cadets would hang out with them on the bleachers for the promise of some of my mom's great cooking.

. . .

HAVING my brother at NMMI with me made my last two years there even better than they would have been, and it has allowed us to share an experience that few people get to have or understand.

40

PIZZA RACES

Old cadets were allowed a lot more privileges and freedom than new cadets. You could order pizza and take it to your room. If you received a care package from home, you got to open it on your own and not have old cadets rummage through it and take your stuff. You could even have a vehicle during your senior year after Christmas if your grades were good enough. It was amazing.

Pizza was considered an essential food group. I'm not referring to the mess hall pizza, which was not even as nutritious and tasty as a cracker with ketchup. I'm talking about ordering pizza from Domino's or actually sitting down at Pizza Hut. I don't like pizza very much anymore, but in the days when my body would forgive my transgressions, I could put away a substantial amount.

"Want to share a pizza?" I asked Drew.

He lived two doors down my junior year on the second stoop of Golf Troop. We were sitting in my room playing cards on a Saturday afternoon.

"You'll never beat me again," Drew said. "Yeah, let's get one."

I smacked my floor with the butt of my rifle and a new cadet showed up post haste. "Excuse me Corporal, Cadet-Recruit Brown, A.

F. requests permission to speak," RAT Brown said as he stood at my door.

"We want a pizza. Pepperoni and mushrooms," I ordered as I gave Brown some money.

"Excuse me, Corporal, Cadet-Recruit Brown, A. F., requests permission to get you some pizza."

"Rest," I answered, releasing him to do our bidding.

"Make sure you time them, maggot," Drew said as Brown was leaving.

Domino's had a "thirty minutes or its free" policy at that time, and cadets held them to it like it was a legally binding contract pored over by lawyers and enforced by the FBI.

When Brown got back with the pizza, we put it on the desk, lifted the lid, and both sat back with our hands on our laps in the starting position. This was a well-loved, well-played, and well-known game among me and my buddies. Each party paid for an equal share of the Italian delicacy, but the fastest eater got the most, maximizing his dollar.

"Go," Drew said.

We attacked the pizza like sharks on a guppy. Each of us grabbed two pieces, folding them over so the crust made the bread of our triangular Italian pizza sandwiches, and we commenced to eating. My chewing was not nearly as on-point as my swallowing so some of the pepperoni passed to my stomach without the benefit of mastication, but I was reaching for the next piece before Drew. As per the rules, you could only take two pieces on the first grab. No pizza sandwich after the first one. There were eight pieces to be had, so I was well into number three before Drew was getting done with his first wedge sandwich.

My mouth was burned by Domino's competence in delivering a piping hot pizza and Brown's swift feet, but I was not going to let a future blister impede me as I ate through number four. Drew was just getting into his third, and the fifth piece was mine. I was the winner. It was times like this that built confidence and conviction in a young soldier, moving on from these victories with the belief instilled by such conquests.

Another favorite pastime my fellow pizza lovers and I enjoyed was to go and hang out at Pizza Hut. It was standard practice that cadets did not go into town alone. The boys in Roswell did not think much of having an additional 700 or so teenage men imported every school year, but the girls were for it in a big way. If you have seen the movie, *An Officer and a Gentleman* with Richard Gere, you have an idea of what things were like. Anyway, if time allowed, a couple of us would go to the local Pizza Hut around lunchtime and order a Coke.

The waitresses were happy as can be to have some young flirty men in uniform, and we were ready to enjoy a full lunch for the price of a Coke. I don't remember free refills being a thing in those days. From memory, you had to pay for each new Coke, but I may stand to be corrected on that. We would sit and wait for the waitress to bring us pizza from the remains of whatever pizzas other guests had ordered. We would watch and hope that those around us would order the toppings we liked best. With luck, they would be full enough to not finish the pizza and hurried enough to leave some pieces behind for our vulture-like nature. A cadet could get more pizza than he could eat if the place was busy and he wasn't overly picky about toppings.

The thought of consuming second-hand pizza now would keep me from enjoying it. But when you had been living off the offerings from the Den of Culinary Atrocities, you could find cold pizza that had previously been on the table and in the care of others to be quite delicious.

41

OLD CADET LIFE

"It's something just to step on these old-cadet corners, ain't it?" Pvt. Harris said to me as we were walking through The Box toward class.

"I know," I answered. "We've only been back a week, but I sure do like being able to walk everywhere, use Little Sally Port, and eat like a human."

"Humans wouldn't eat that food, though."

I had to agree. "How are the RATs in Alpha Troop?" I asked.

"Probably as dumb as the ones in Golf," he said with a smile. "I hear your brother's a RAT."

"Yeah. I think he's doing pretty good so far. Survived Hell Week with flying colors and my buddies over there tell me he's a pretty good RAT."

I had been moved to Golf Troop and my brother was doing his RAT year in Delta Troop, which had been my RAT troop, as well. Being an old cadet was so much better than the life of a RAT, and life was good. I was enjoying rodeo, hadn't been in much trouble yet, and the classes weren't boring. Having a challenging course load was not appreciated at the time, but it helped keep you on track and set you up for future successes in a way that few other things can.

"Is he a cowboy, too?" he asked.

"Little Mike's a magician with a rope. His main event is team roping, and he does it very well."

"Great," Pvt. Harris said. "I'll look out for him if I see him."

"Thanks bro. Have a good'un."

We slapped hands and parted ways as he headed to his class, and I headed to Major Bartl's English class. It was so cool to have my little brother in the Corps of Cadets. I got to see him every day at the rodeo grounds. He would spin steers for me, rarely missing, even though I didn't hold up my end of it very well on the heeling side. Team roping did not seem to be a natural to me, but it was something to see Little Mike in action.

As the year wore on, I found another special spot for myself in the NMMI Motor Pool. The garage was located behind the rodeo grounds, and most cadets had no idea that it was even there.

"Hey there, my name's Chris," I said as I walked into the mechanic bay where a man was working under the hood of a school truck.

He came out from under the hood, "My name's Tom," he said.

Given a minute or two and I would have known it from the patch on his grey work shirt.

"I'm kind of interested in mechanic work," I said. "Mind if I hang out?"

He looked at me a little sidewise. It must have been the first time a cadet had made that sort of request. It might have been the first time a cadet had ever walked into the motor pool for that matter. "All right with me, but Smitty will have to approve it. HEY, SMITTY," he hollered.

An older man came out of an office at the end of the three-bay garage. The name on his shirt was also Tom, but he obviously answered to Smitty.

"This is Chris. He wants to hang out in the motor pool with us. What do you think about that?"

"No one's ever done that," Smitty said. "You'll have to get Col. Smith to sign off on it, I reckon."

Col. Smith oversaw the motor pool from a bureaucratic perspective, but I'm not certain he knew where it was any more than most of the cadets.

"OK," I said. "I'll find him."

I left the motor pool and headed back toward The Box. Finding Col. Smith was a little harder than I had expected. Within a week, I met with Col. Smith and was able to get permission to learn some stuff at the motor pool, even though he thought it was an odd request. Mechanic work interested me, but I hadn't had enough experience to find out that I really don't like doing mechanic work. I know it now.

"You get permission?" Tom asked.

"Yes sir," I said. I handed him the note that Col. Smith had signed for me.

"Perfect," he said. "You're basically going to learn how to change oil."

"And wash windows," Smitty said from inside the van he was washing windows in. "Grab a roll of paper towels and get started."

I went to work, and we had the windows done in no time. Smitty pulled the van out of the bay, and we went into the little office to have a break. The office was comfortable with a small desk, some shelves filled with parts and maintenance manuals, a table with a coffeepot, and about five or six office chairs around a little coffee table. It seemed like a lot of chairs for a small spot to me, but I didn't quite yet know about the life of the motor pool.

We sat down and Tom handed me a cup of coffee as we got to know each other. I wasn't in the habit of afternoon coffee anymore, but I took it to be polite. I looked around for some Jim Beam since afternoon coffee with Pony had always had the extra kick.

Smitty had been at the motor pool since time immemorial. His real name was Tom Barley, but he had been called Smitty since he was a kid. Though I asked, he claimed the nickname did not come with a story, but it fit him way better than Tom did. The other Tom looked like a Tom, so I left him alone and let him keep that name. He had been a mechanic at a dealership in Dallas and had moved to Roswell when his wife had gotten a job there.

"This is the easiest job I've ever had," he said over the rim of his coffee cup. "Twenty-seven vehicles and we change the oil and put air in the tires. Any more than that and it goes to a garage in town."

"And wash the windows," Smitty added. It was his thing.

"What do you do with the rest of your time?" I asked.

"What we're doing right now, boy," Smitty answered from behind the desk.

"Anyone home?" a voice hollered from the bay.

"In here," Tom yelled back.

One of the campus cops came in and helped himself to a cup of coffee and took a seat with the familiarity of one who had been there many times before.

"Jack, this is Chris," Tom offered. "He's here to learn some mechanic-ing off'a me and Smitty."

"You ain't going to learn that here," Jack laughed, looking at me. "Where do you live?"

I was in my civies instead of a uniform, so Jack didn't know that I was a cadet. "In Golf Troop," I said.

"A cadet? In the motor pool? Does anyone know you're here?" Jack asked.

"Yes sir, yes sir, yes sir," I answered.

"Well, that's a new one. Ain't that a new one Smitty?" Jack asked.

"Sure enough," Smitty agreed.

Pretty soon another campus cop came in and we talked, some chewed tobacco, some smoked, and I looked around and felt like I was right back in the middle school boiler room with my janitor friends; the guys who knew where everything was, what everyone was doing, and what was really going on behind the scenes.

I hung out at the motor pool whenever I could. Jack was right about my not learning too much about being a mechanic, but I did make some great contacts and had a special place to hang out and feel like a civilian from time to time. Those contacts served me well later on in my cadet career.

42

PLAYING PULL-OFF

By my senior year, I was the Mounted Color Guard commander. Rodeo and horses interested me a lot more than things military. In the corps, I was living in India Troop, but my formation was with the regular Color Guard. These were dismounted cadets who raised the flag in the morning, lowered it at night, and shot the cannon. Since I was technically on the Mounted Color Guard, I didn't have to be there for the dismounted Color Guard every formation, so life was easy for me. With the assurance and confidence that only a teenager can possess, I declared my title as Mounted Color Guard Representative to the Dismounted Color Guard, a previously unknown position that allowed more freedom than the military would wish upon a lowly sergeant. I could sleep in past the blast of the morning trumpets and I didn't have to go to SRC with my troop. I went with the dismounted Color Guard about twice a week and was living as unrestricted a life as was possible for any cadet. Again, I was finding that special place, just like the janitor's area in middle school and the motor pool at NMMI.

When it was time for an inspection and parade, we of course had the perceived massive and lengthy job of getting the horses ready for their performance. None of the brass understood anything about horses and didn't know that we could have easily gone to the arena and been saddled and ready in short order. Not that any MCG cadet would have ever let them know. By tradition, this had always allowed the Mounted Color Guard guys to miss the inspection altogether, and we would get out to the arena in the morning and have a full day of looking busy while goofing off. The knack of looking like you are accomplishing something while you are slacking off is perhaps the most valuable soldier skill. That is why a lot of second lieutenants carry clipboards. Problems arise when you are caught truly goofing off without the appearance of progress.

"Thanks for getting me out of the inspection," my buddy Sean said from the top rail of the arena fence. "I hate those things."

"Everyone hates them," I agreed.

He was a bronc rider, so he was a member of the NMMI rodeo team but not on the Mounted Color Guard. About ten of us were sitting on our horses just inside the arena, getting ready to play one of our favorite arena games that we called pull-off. The game was simple and based on the popular freeze tag that kids used to play at recess or in progressive PE programs. There were two teams, and the object of the game was to pull the other riders from their mounts. Once you were pulled off your horse, you could remount, but you had to sit there until another member of your team touched you so that you could rejoin the fray and start pulling people off their horses again.

"Hey man, no problem," I responded. "Of course, the MCG needs help getting ready. You're on my team if you want to ride Shorty."

He went to get my other horse, but I was on Pelvis. While Pelvis could be a little animated and not necessarily the best roping horse, he was one of the best pull-off horses you could have. He was massive and didn't mind running over and through other horses and mounted cowboys. I think he would have been a great jousting mount if I had been born in another time and had been a knight.

Sean was strong and brave, so he made a great addition to anyone's pull-off team. Shorty was a great heeling horse but lacked the size and strength for playing pull-off. Sean had to make up the difference. Once the teams were decided, pandemonium commenced.

I rode Pelvis with the intention of being like a bowling ball. It was great to hit another horse in T-bone fashion, make that horse stumble, and grab the other cowboy and pull him to the ground. The downside was that when I got pulled off Pelvis myself, it was a long way to fall and harder to get back on bareback in the middle of the arena.

I smashed hard into my brother's mount and tried to grab the collar of his shirt. He ducked, and Leon pushed my out-of-balance self over the side of Pelvis. I landed hard amidst a mass of stomping hooves and did my best to keep my reins. If your horse ran off, that was the worst. You had to stand there while a teammate retrieved your horse. Being without a horse in the middle of all the other mounted cowboys could be a challenge. They weren't too interested or concerned about your safety, so you had to take that responsibility upon yourself.

I popped up and remounted as the main part of the skirmish moved away. I had to sit still and wait to be tagged back in.

"They got me," I yelled to any teammate that might be able to unfreeze me.

"I'm coming," I heard Dick respond.

Dick rode a big paint horse, and he was galloping toward me when Kevin attacked and they both ended up in a heap on the ground. It wasn't looking good for my team. Little Mike had pulled Sean off in another corner, and it was just about over for my guys when I heard an angry yell from the gate.

"EXACTLY WHAT IS GOING ON HERE?" the man yelled. "WHO IS IN CHARGE?"

Crap. I looked, and it was Col. Bruno. He was the officer in charge of all the TAC officers. These were the active-duty Army officers who controlled the military interests of the NMMI Corps of Cadets. Col. Bruno had been in Delta Force in Vietnam and had very little patience for any cadet who was not inspired to become the best soldier that he could be. I was not in the category of aspiring super-soldier. That he could make my life miserable was a given, and I just knew that my short time as MCG Commander had just come to an end.

All eyes were on me as I rode Pelvis toward the waiting colonel. He was in his dress uniform for inspection, and how he had come to the arena is still a mystery, but there he was.

"I am," I said. To Pelvis's credit, he was walking slowly and with his head down, every bit as meek as I was feeling.

"Just exactly what is going on here, Sgt. Gregory?" he asked.

It was not a great honor that he knew my name by sight, even though I was not in uniform. My undesired participation in tour squad had led to that familiarity. It also led to him expecting the worst from me.

"We are getting ready for the parade, sir," I answered.

"There are only six horses in the Mounted Color Guard, and I see at least a dozen horses and cadets out here. You have some serious explaining to do. I will see you in my office after BRC at 0730 tomorrow. Do you understand?"

"Yes sir."

"This had better be the best parade of your life," he said to me. "Don't be late tomorrow."

"Yes sir."

"You had better have a very good explanation for this. There is an inspection going on." He seemed at a loss, but I could tell he wanted to continue and ream me out right there.

"Yes sir." What else could I say?

"I can't believe what I am seeing out here. You will be at my office tomorrow at 0730."

"Yes sir," I seemed to be stuck in that groove.

With that, Col. Bruno executed an about-face, and marched through the stall area and back toward The Box where the inspection was getting underway. If people really did produce smoke from their ears, he would have been leaving a trail like a steam engine. We still had about three or four hours until the parade.

"Oh man," Little Mike said, riding up on Corky. "You're in for it, bro."

"Probably," I agreed.

"What are you going to do?" Dick asked.

"Nothing to do now," I said. "At least I have the night to work on it."

"Glad it's you and not me," Sean added.

"Well men, we might as well finish the game," I turned and started loping toward the end of the arena.

"Hold on," Kevin hollered. "You guys are all already frozen. You can't just start over like that."

"New game," I yelled back and started toward him on my charger with the full intent of pulling him off. He evaded and the game started fresh as we tried to kill each other without dying. Amazing that we survived that game, but it sure was fun.

The parade went off without a hitch and we all looked like mounted heroes as we rode through the mass of marching cadets, each flag held straight, each horse walking and turning in formation with precision and grace.

"OK guys, just let me do the talking," I said to my guys who were waiting outside Col. Bruno's office the next morning.

All my guys had decided to come with me. They figured it was going to be quite the show, and the chances of getting hit by friendly fire were small since it all rested on the commander's shoulders.

I had experience with being in trouble and experience with talking my way out of it, as well. My youth had provided a lot of practice with that sort of thing. This one was going to be a sticky wicket, as they say, but there wasn't any choice.

"Col. Bruno, Cadet Sgt. Gregory, C. L., reporting as ordered," I said as I stood at attention, my right hand held stiff in salute, my eyes boring a hole through the wall above his head where he sat at his desk. My fellow cowboy cadets stood at attention behind me.

"At ease," Col. Bruno said as he returned my salute. "Exactly what in the Sam Hill was going on out there yesterday?" he asked.

"Sir, we were getting the horses ready for the parade, sir," I answered.

"That didn't look like any parade preparation to me," Col. Bruno said.

"Do you remember when I got bucked off in the parade my RAT year, sir?" I asked.

"Yes, so what?"

"Well sir, if my horse had been prepared the way I was getting those horses ready, he would never have bucked, sir," I answered. "I learned a lot from that experience, sir."

"How was running around the arena and hitting each other helping?" he asked.

My story was thin, but it was all the substance I had, and I wasn't going to abandon it. "Well sir, those horses are athletes that get very pent up if they aren't properly warmed up and worked prior to having to stand out there in the parade with so much going on around them, sir. What you witnessed, sir, was my way, as Mounted Color Guard Commander, of ensuring that the horses have been exposed to everything I can expose them to prior to being in the parade, sir."

"And why did you need so many people out there when there are only six horses in the MCG?" he asked.

A valid question, but I had another answer. "Well, sir, we needed the extra riders to make sure the horses that were in the parade were well prepared prior to the parade, sir," I answered. "Plus, sir, the extra manpower helps us make sure that the horses are groomed and that the leg wraps are correct before we go out on parade, sir."

"I'm not sure I am buying all this, Sgt. Gregory."

"Well, sir, no one has been bucked off in a parade since I took command, sir. So, sir, something must be working, sir."

No one had been bucked off in a parade since I had been dumped by Pelvis in my RAT year, and I had never heard of any prior, but that was beside the point. We were all on seasoned rodeo horses, so the reality was that these horses had already been places and seen things, so the repeat of my embarrassing incident was unlikely anyway. Not that I was going to share that with Col. Bruno.

"Get out of here," Col. Bruno growled. "See if you can do it with fewer cadets missing the inspection next time. Dismissed."

With that, we all left and headed to the PX to have a coffee or milkshake, depending on our personal bent, and discuss the great escape that they had just witnessed.

"Congratulations," my brother said. "You are unbelievable."

"Well, lucky for us, I am just believable enough," I answered, putting my arm around Little Mike's shoulders.

We all laughed. It was a lucky break, but unfortunately, future parades did not include a big game of pull-off to get the horses ready. We also had to leave our rough-stock buddies to the unhappiness of military inspections.

43

NMMI FARRIER

There was a farrier who shod horses for Joel Webb, whom we called Boss. He was our rodeo coach and a great man and mentor of my youth. This led to him getting more business from the cadet rodeo cowboys, so he shod my horses as well. His name was Rod, and I remember him as being a very grumpy guy. Seeing that I was inquisitive, my incessant questions might have been annoying to him, and he may have been much happier when I wasn't around. Hard to say since it is impossible to judge what he was like when I wasn't around him since I wasn't around him to judge, if you see what I mean. The other NMMI cowboys weren't as interested in farriery as I was, so I'm sure that their experience with Rod's personality was more favorable than mine.

"What do you have that for?" I asked Rod, referring to the forge that was sitting on its back in the bed of his truck. The firebox was full of crushed nail boxes and there wasn't a propane bottle to go with it.

"That's a forge for when I need something special," Rod said. "Hand me those nails." He was under a horse and getting ready to nail up.

"I haven't ever seen you use it."

"I don't use it very often. Only when I really need to hot shoe a horse or make a shoe or something like that."

"Can you use it when you shoe my horse?" I asked. I was eager to see a horseshoer do something with a hot shoe or make a shoe, which I had never seen before.

"I get an extra $20 when I hot shoe," he said.

Bear in mind that shoeing a horse at that time was only $25, so the added $20 was substantial.

"Wow, is it that good for the horse that it's worth the extra?"

"No, I don't think so," Rod answered.

"Then why do you have the forge?"

"I told you, when I need something special."

"I don't understand," I said. "What is something special?"

"Why don't you go rope or something?" he asked.

His annoyance was showing, and from this distant perspective, I can certainly understand it. Rod would have made a politician, for he had mastered the art of answering a question without imparting any information whatsoever. Whether he knew anything worth knowing or not, I still don't know. My horse's shoes stayed on, and the horse stayed sound, so he fulfilled the basic requirements of being a horseshoer. If he had been a little smarter, he could have had free and willing help.

I was looking for a new heel horse, and Rod got wind of it. "I hear you're looking for a horse," he said one afternoon as he came to find me. He was friendlier than he had ever been.

"Yes sir," I said. "I need a new heel horse."

"I know a guy that's got a great one if you want me to set it up so you can try it," he said.

"Sounds good."

"Great. I'll have him bring him around on Friday and you can practice on him. Good feet. Nice little horse," Rod said.

"How much does he want for him?" I asked.

"Well, he is a very finished heel horse, but I think he would take $1750 for him."

"That's a little much for me," I said. "Better not bring him by."

"You don't have to buy him if you don't like him," Rod said.

"Yeah, but still, I can't spend that much on a horse anyway."

"You should just try him," Rod said.

"OK," I agreed. "But I'm not in the market for a $1750 horse"

The man brought the horse and dropped it off in a stall. I roped on it, and it was a decent little gelding. Nothing special, and certainly not worth $1750 in 1986 dollars. Rod came to that practice to try and persuade me to buy it. By the end of practice, I was certain that this was not the horse for me, and I let Rod know.

"Dang kids," Rod said. "You don't know a good deal when it's right in front of you."

"I told you I didn't want a $1750 horse," I said.

"You don't even know what a horse is worth then," Rod told me. "How about $1500."

"Sorry," I said. "That is not the horse for me."

Rod grumbled and drove away. The owner of the horse came to get it as I was headed back to the barracks in uniform. I saw him loading the horse and walked over to say hello. Being in uniform, he had no idea that I was even a cowboy or part of the rodeo team.

"Good afternoon," I said. "Is that your horse? My name's Chris."

"Yeah," the man said shaking my hand. "I'm Cotton. My horseshoer thought he could sell it to one of the cadets, but then he called my wife and said the boy didn't want it."

The man didn't know that I was that cadet, and I didn't tell him.

"That's too bad. What does this horse do?" I asked.

"He's just a heel horse," Cotton said. "I've been trying to sell him for a bit but haven't got him moved."

"How much are you asking for him?" I asked.

"$800," Cotton answered.

Well, that was much more like it. That was well within my budget and the horse was worth that much.

"I was that cadet," I said. "Rod told me that you wanted $1750 for this horse."

The man looked surprised when I said that. I could see his gears start to turn as he put two and two together and he smiled at what Rod had done.

"I guess he was planning on making a little commission," Cotton said.

"Yeah, a little commission," I agreed.

"Do you want the horse?" he asked.

"I would buy him for $800."

"You got a deal."

He unloaded the horse, and I gave him a check. We shook hands, and I led the horse back into the stables.

About a month later, Rod was shoeing in the stables, and I walked up. "Hey Rod, can you shoe a couple of horses for me?"

"Yeah," he answered. "I'll get them next week. Do you have two to shoe? Did you get another one?"

"Yes sir," I said. "I bought that little horse that I tried last month."

Rod stood up. "Cotton's horse?" he asked.

"Yes sir."

"I thought you didn't want that horse," Rod said.

"Well, I didn't want him for $1750 or $1500. Turns out he is perfect for $800."

"&^%t," Rod said. "You'll need to find someone else to shoe him."

"Sounds good," I agreed.

I ended up with another farrier for the last shoeing of my senior year at NMMI. I don't remember who, but I had learned some great lessons. Don't believe every salesman, and if you are a farrier, don't get involved with horse-trading. It is a line I still don't cross to this day. I'll put owners and buyers in touch if I know of them, but I sure don't get involved with any part of the deal. I would rather have the customer than the commission.

44

BULLDOGGING SCHOOL

"This is such a great horse," I said to my friend Jay Wally.

"Thanks, man," Jay said. "Glad you like him."

I was riding Jay's big bay doggin' horse, Texaco, because I didn't own my own bulldogging horse at that point. In steer wrestling, using someone else's horse is called being mounted. Jay lived on the reservation in Crown Point, New Mexico, and he and his brother were at all the high school rodeos and willing to mount folks who wanted to ride their horses. If the cowboy is successful, the owner of the horse and the hazer get a part of the winnings. The hazer is the cowboy who rides alongside the steer to keep it running straight.

Texaco and I did well together. I always liked a horse that was a bit bigger because I felt it could really stay in position and seemed to throw my feet out in front of a steer better than a shorter horse. It was further to get down to the steer, but I still felt the bigger horse fit my style better.

"How would you like to go to a bulldogging school in Kansas over Thanksgiving?" I asked Jay.

"Sounds good," he said. "Let's make a plan." This was a lot of words for Jay to put together at one time. He was much better at sign language and said more with gestures and facial expressions than he did with words.

We left New Mexico as soon as school would let us and headed east with Dad's truck and trailer, Texaco blanketed against the cold in the back. It was going to be an epic adventure for Jay and me; three days of nothing but throwing steers and having fun.

The year was 1986, and the winters of the Midwest were much different from what Jay, Texaco, and I had ever experienced in the southwest. Getting out of the truck in eastern Kansas, I got a freezy headache just walking a hundred feet to the barn, an icy remnant of snow crunching under my boots.

Jay said, "I have never felt a cold like this before." He continued with some Navajo, and it was more than the "Yutta-hey," I was used to hearing when he greeted me. He was probably cursing the cold in his native tongue.

"Me neither," I agreed. "Wish I could cuss in Navajo."

He smiled as we put Texaco in the stall. We dumped the five-gallon ice cube from his bucket and replaced it with water that was not yet solid. The hotel was warm and the night uneventful, but it was hard to sleep in anticipation of the class to come.

The class was about technique and volume, just what I was hoping. The owner of the school had made a dummy that could be hauled with a four-wheeler, and we practiced riding by and jumping off the horses as we worked on becoming better steer wrestlers. We did a lot of chute dogging, which is getting beside the steer in a bucking chute and running with it to get in a position to throw it down. The afternoons ended with jumping live steers and making runs. It was phenomenal, and Jay and I could feel our skills advancing by the hour. Though I was fit and young, I remember being as sore as I had ever been by the end of the three days. There was no need to apply ice since the exposure one got by walking even ten feet outside was sure to freeze your aching anatomy.

I was offered a chance to return to Kansas and stay in what passed for an office in the indoor arena. It was primitive, but at least had some of the basic requirements for humans to exist. I took it, thinking that this was going to be the catalyst to advance my desired career. At that time in life, my plan was to just be a rodeo cowboy and go to the National Finals Rodeo (NFR) every year. Of course, in my dreams, this would lead to fame and fortune. God laughed.

Jay and I returned to New Mexico, he to the northwestern reservation, me to the grinds of NMMI. I had a great rodeo season that year and tied with another NMMI rodeo cowboy for the state champion steer wrestler. The saddle winner at the end of the year was determined by a coin flip, which he won. I did get the buckle.

"Mom and I are sure proud of you," my dad said. We were at the Chevrolet dealership in Las Vegas, New Mexico, and by all accounts, it looked like I was going to be getting out of NMMI with a diploma and on time. "A man needs good transportation and responsibility. This truck will be both."

Little Mike and I were on Christmas break from NMMI. Dad had set up the truck deal for me before we came home for the holiday, and I was going to be able to take the truck back to NMMI for my last semester.

My folks had graciously paid the lion's share of the new single-cab half-ton pickup truck. I was left with a small payment every month, which was the responsibility part of it. The success that my folks had achieved for themselves could have paid for the whole truck, but their knowledge of raising young men meant that I had to have a part in the payments, as well.

"Thank you so much," I said. "This will be a great rodeo rig."

I had found and bought a two-horse, goose-neck trailer with a tack room in the front of it. I was now ready to hit the road and live my rodeo dream of throwing down steers from coast to coast. After school, of course.

"Take care of it, and it'll take care of you," Dad advised.

His council was sound. My application of it was not, but I'll come to that soon.

45

SHOOTING RABBITS

My dad had a good friend and hunting buddy, named Loren, in Roswell who had some land and horses just outside of town. Since I had a truck for the last semester of my senior year, I got to get off campus a lot more than I ever had before.

There were so many rabbits in the Roswell area at that time that they were considered a nuisance.

"We was thinking about spotlighting tonight," I said to Loren. "Do you have an extra rifle Dan can use?"

Loren smiled. "Only about thirty of them. What do you want?"

"Just a .22," Dan said.

I had my little .22 Remington that stayed behind the bench seat in a scabbard in my truck. There were gun racks in the back window, but I only put extra rifles in those and kept the little Remington behind my seat. There weren't any rules against a boy having a rifle in his truck in those days. Different world.

Dan and I drove through pecan orchards and past fields that were covered with rabbits. It was a good night for hunting, but time got away from us.

"We're late," Dan said. "You'd better turn your lights off when you pull into the parking lot."

I agreed and slowly drove my truck into the NMMI parking lot that sat between The Box and the parade field. It was thirty minutes after Taps and curfew, and we were going to be lucky to be in our bunks in time for bed check.

"So far so good," I said, turning off my truck.

Dan and I got out of the truck and walked in the shadows toward The Box.

"Who's there?" a voice asked from behind us. The beam of a flashlight sent our silhouettes to the bricks of The Box that was just an agonizing few yards away.

"Cadet Holland and Cadet Gregory," Dan answered. He had raised his hands like we were being held up by a mugger.

The campus cop walked around us.

"Is that you, Chris?" Jack asked, his flashlight shining bright in my face.

"Hi, Jack. Dan and I have been spotlighting rabbits. Time got away from me. How's Brenda?" I finished with an obvious ploy to direct the conversation and remind Jack of our times together in the motor pool.

"What troop are you in?" he asked Dan.

"We are both in India," I said.

"Well, come with me and I'll escort you back to your rooms so the TAC officers don't freak out," Jack said. "Have you been drinking? You wouldn't believe what Brenda's sister did now," he finished.

"We had a beer or two," I admitted. "What did she do?"

"Well, at least you're walkin' straight," Jack said. "Stay close in case there is a TAC on the stoop."

We stayed close to Jack as he rounded the corner and started up the steps to our stoop in India Troop. "My dang sister-in-law left her husband and kid in Phoenix and wants to come live with us now," Jack said. "That gal ain't nuthin' but trouble."

"Man, I'm sorry to hear that," I said. "Thanks a bunch, man. I'll see you at the motor pool next week if you're back on days."

"Yeah, should be. You boys be careful and get to bed now."

With that, Jack followed his flashlight beam back down the stairs. I stepped into Dan's room to say goodnight.

"Dude, you're unbelievable," the quiet Dan said. "How did you just pull that off?"

"Remember my friend, it really is all about who you know." With that, we bumped fists and I headed to bed.

46

GREEN TAIL ROPE COMPANY

"I can't believe how many ropes I'm going through," I said to Little Mike. "They keep losing their feel and the hondas turn."

"You and me both," he said.

"I'm going to start a rope company," I said.

Little Mike rolled his eyes, a common response from many of the people who knew me back in those days. Luckily, nature had made me immune, and I was starting to take that look as one of compliance and confidence in my many schemes. The entrepreneurial spirit has always run deep in me, and this was just one more of those stops along the way that eventually led to being a farrier and having a school.

"How are you going to do that?" he asked.

"I don't know yet, but someone has done it before. How hard can it be?"

"Yeah," Little Mike said. "How hard can it be?"

I found a guy in Albuquerque who made calf ropes. He gave me a little advice on making the hondas, which are the knotted loops on the end of a lasso, since I bought ten of his ropes.

He described the honda-tightening contraption I would need. For those aspiring entrepreneurs among you, I'll describe it here if you want to make one.

This contraption was a ten-foot-long metal table with a winch bolted to one end of it and a wheel rim from a tractor tire welded to the opposite end. Between the winch and the rim were two tracks that a thick metal plate traveled on. The plate was attached to the winch cable and had a three-inch-long post welded to it that stuck straight up. Tied ropes were placed on the rim, hondas on the post, and the winch cranked to tighten the knot. It turned out to be a very slow process.

The method of making a lariat was more involved than one might imagine, especially if that one is young and doesn't yet know that there is always more to any business than meets the eye. I began by getting six-hundred-foot spools of the nylon rope and stretching it out across the pasture. Each end was tied to a softer rope that was used as a tow rope in water skiing, and the kinks in the nylon rope would end up twisting into the tow rope. I had about ten spools of rope tied to a pipe that was chained to a tree, with the other end of the ropes tied to a pipe attached to a truck. It was quite an event just to put that together. Luckily, we didn't have any cattle in the big field at the ranch, so my rope-stretching spot was not a problem.

My kind-of mentor had suggested leaving them stretched out like that for at least two weeks, but of course, I did not have that kind of time. By the end of the first week, I was cutting the first coil up and tying hondas. Once the hondas were tied, I sewed a green leather burner into the end of the honda. Burners are used to protect the nylon rope from friction where the rope passes through the honda. When an animal is roped, there can be a lot of damage to the honda if it is not protected by a burner. Then I would dye the tail of the rope green and have a finished product to sell.

"Looks like a lot of work," Little Mike offered as he watched me.

"Well, we'll have a lot of ropes to use, and I should end up making a lot of money on these others."

He picked one up and gave it a few expert swings.

He said, "This one feels good. I'll give it a go at Boyd's tonight."

"I've about got a dozen or so done," I said. "Maybe I'll get them all sold and we can rope for free."

Funny how the mind does that. Selling ropes or not has nothing to do with the cost of roping. The same as shoeing horses at rodeos does not make rodeoing free. Don't tell that to any youngsters learning how to rationalize their desires and expenses. That rationalization keeps the dreams a-comin'.

We went to the roping, and I sold about half of the ropes I had with me to Boyd. They were 7/16 full, thirty-foot head ropes. A bit heavy for some, but I knew Boyd liked that size, so that was why I had gone with that spool first. Little Mike used his and did well, but he is the kind of roper who makes the horse and rope look way better than they are, kind of like the pool player who can beat you with a broomstick instead of a pool cue.

"My rope turned," Little Mike told me when we got back in the truck.

"You're kidding?" I asked. "You only roped about five steers."

"I know, but there it is. The honda is rolled right around."

There was no denying it. Little Mike's new Green Tail Rope had a twist in it.

"Sure hope that's a fluke," Little Mike said.

"Me, too."

It wasn't. Since the ropes had not had enough time to get the kinks out, about half of them would roll pretty much after roping just a few steers. I took all these back for a full refund.

"Any ideas?" I asked my brother. We were looking at the pile of returns.

Little Mike said. "Why don't you tie a honda in the other end and stretch them again individually? Just cut off the bad side."

It was a great idea. I tied a honda in the opposite end and stretched each rope between a corner post and my pickup. I didn't have to use the honda tightening machine since the truck was pulling hard enough.

The result was a pile of ropes that were about a foot less than the desired thirty or thirty-five feet. Most of them ended up with a honda that didn't roll, and the guys that ended up with one of those ropes seemed to like them well enough. For me, I had put in a lot of work and effort to end up with a business that was not going to make much money compared to the amount of time needed to run it. Green Tail Rope Company became a statistic. It also became one of those valuable life lessons and failures we all must have if we are going to learn and someday succeed.

47

SPRING BREAK IN PURGATORY

Skiing is a skill best learned young. The body recovers from injury with ease, and the brain is still brave enough to let that body get injured, so a young skier can have a lot of fun letting it all hang out. I had been in ski clubs throughout public school and had a lot of fun skiing as a kid. During spring break of 1986, Mike, Nick, and I went to Durango, Colorado, to have fun skiing Purgatory. Nick was a bronc rider on the rodeo team, and we always had a lot of fun together.

Computers were just coming out in the 1980s, and most people didn't have access to their own computers or printers. Luckily, there were some affluent cadets whose parents had provided them with cutting-edge technology. Like all technology, it can be used for both good and evil, and I had a friend in Bravo Troop who was making and selling fake IDs. I had paid him $10 for a West Virginia driver's license with my photo and the name of Robert Martin. It was pretty good for a cardboard fake, but it was better if it was left in a plastic sleeve in the wallet. That meant you had to hand your wallet over when you used it.

"I need two six packs of Miller High Life," I said to the man behind the window of the drive-through liquor store on the main drag in Durango.

"Do you have an ID?" he asked.

Smiling, I handed over my wallet with the never-fail West Virginia driver's license.

"West Virginia," the man said. "I've never seen one from there before. Hold on."

With that, he opened his handy ID identification booklet and found West Virginia. He showed the page to me through the window, and my ID looked nothing like it. At all.

"You better come on in and wait for the cops," he said.

I pulled forward to park the truck. He had my wallet with my real IDs and whatever money I had, so I didn't have a choice. Little Mike and Nick were shaking their heads at me, the "I told you so," look in their eyes, if not their words.

I went into the liquor shop to wait for the cops. The man was very nice, and we had a civil conversation while I waited. By the time the cops arrived, he was regretting not just selling me the beer and letting me enjoy the rest of my spring break. "Just so you know, he was a polite young man," he informed the heavyset blond cop who came through the door, donut sprinkles stuck to his chin.

"I'll take it from here," the cop said. He read me my rights, handcuffed me, and led me outside to his patrol car, not gently.

Little Mike and Nick were in my truck talking to the other officer. He was the poster of what a cop should look like and the opposite of the chunky blond who had done all the hard work of arresting me. His dark hair was perfectly trimmed, his mustache symmetrically grown with military precision and order, tanned skin despite the snow on the ground, and muscles straining the short sleeves of his uniform shirt. You could tell that he wore short sleeves year-round to make sure his muscles received their due admiration.

Later, Little Mike would tell the story, in his special way, of this Rambo cop marching up to the truck that he and Nick were sitting in. He had taken off his aviator's sunglasses and tapped on the window. Little Mike rolled it down.

"Excuse me," the burly cop said in a voice that could have belonged to a teenage girl or a frightened little boy. "Are you with him?" he asked, pointing toward my handcuffed form being led to the cop car.

Stifling a giggle, Little Mike confirmed our association with a nod.

"Please follow us to the station," the man said in his dainty tone.

As we drove to the station, the chunky cop asked, "Do you go to NMMI?"

I had been wearing my NMMI Letterman's jacket, so either I was from there or dating someone that was.

"Yes sir," I said from the backseat. "We're on spring break. It's my senior year."

"How long have you been going there?"

"This is my third year," I answered.

"Is it tough? I heard it was a super hard school."

"Yes sir," I said. "The academic side is pretty hard, but the military side is not too bad once you get past your RAT year."

"My nephew went there last year," he said.

Maybe this could lead to a connection that might just save my bacon, so I perked up and hoped his nephew was a friend.

"Really? I'll bet I know him. What's his name?"

"Doug is his name, but he only lasted two days. Came home with his head shaved and a bunch of horror stories. You would have thought he had been in the trenches in World War II or something," he answered.

That just went from bad to worse. He had a family member who was too soft for NMMI. Oh well, I might as well sit back and shut up, was all I could think to myself. A night in the cell might not be all that bad. I generally made friends fast enough.

"Sorry," I said. "NMMI isn't for everyone."

"You can say that again," he continued. "My sister spoiled that brat since he was born. They only sent him there as a last hope to make a man out of him."

I sat back up. Maybe it was going to be ok after all.

By the time we made the station, Chunky and I were well on the way to being friends. He discussed letting me go with his partner, but they decided that the least they could do was call my dad. We went into the station, and he took the handcuffs off me. I sat in the chair at the side of his desk.

"Want a coffee or donut or anything?" he asked me as if I were just visiting.

"No, I'm fine," I said.

"I'm sorry, but I have to call your dad. Don't worry though, you won't have to spend the night in jail or anything like that."

I could tell that he felt bad about it and was apologetic. As I've stated, I had experience with being in trouble. This was a case where the person in charge could feel guilty for my being in trouble, and I was going to exploit that if possible.

"Calling him will be worse than a night in jail if I can trade you," I offered.

He smiled sympathetically and said, "Sorry son, this is for the best."

In hindsight, I wish the girly-sounding cop had called because I'm sure my dad would have said "Yes ma'am," at least once. Chunky blond at least sounded like a man. His side of the phone call that I could hear was apologetic as he spoke to my dad. When he handed me the receiver, my dad's side to me was anything but apologetic.

"You get in that truck and head right this way right this minute, young man," my dad said. The anger in his voice came through electronically without hindrance.

"Well, sir," I said. "There is a massive snowstorm between here and home. The roads are closed for right now."

"You're lucky," Dad said.

I had to agree with that, in a way. If I had truly been lucky, we would be sitting around a hot tub with a bunch of open beer bottles between us right now. It is all relative.

"You boys get home as soon as you can."

"Yes sir," I said.

He hung up, and Durango's finest released me to my brother and buddy who were patiently waiting for me. The cops had returned my wallet but kept the fake ID. I was kind of wishing I had asked for it back, but I think that would have been pushing my luck.

We had to resort to the old-fashioned way of buying alcohol for guys who are too young to do so and not in Las Vegas, New Mexico. In Las Vegas, you only had to be old enough to drive and go through the drive-through window. Carding people wasn't really a thing. Durango was a little more modern, so we had to find an adult and get him to buy us some beer. The main problem was that it cost a little more that way. To say that it was a different time would be a gross understatement. We ended up with our beer, and the storm saved us from Dad for a few days. It also gave him time to calm down from being called by the cops.

We were not in a big hurry to get back to Las Vegas. Of course, we wouldn't want to do anything dangerous. Being young rodeo cowboys, safety always came first, and we waited until the roads were super clear before venturing south toward home. Besides, fresh powder in Purgatory and cold beer was hard to pass up. In fact, we made it to the ranch just long enough to get our laundry done before heading back to classes. I don't even remember Dad being all that mad by the time he saw me. Time really does heal all wounds.

Little Mike and I returned to NMMI and finished out the year. Rodeo season was in full swing, and we had a great time in the pursuit of rodeo stardom. It was my last year, but Little Mike still had one to go. I did my first year in NMMI without him by my side, and he was to finish his last year without me being there. My times there with him were better than those without. I don't know if he'd say the same.

48

HOME STRETCH

"Don't you feel like we could be attacked at any time?" I asked Little Mike.

We were riding into the Red Rock Arena in Gallup, New Mexico. He was the header of our team-roping team.

"It is eerie to see the cliffs covered with Indians," he agreed.

"I feel like we are about to live a scene from just about every Western I've ever seen."

He nodded and we rode through the gate. As I remember, that arena had always been good to Little Mike and me and we stretched one out.

Rodeoing in New Mexico in the 1980s was rodeoing at its best. There was a lot of great competition, and the general public looked at rodeo as America's number one sport. But to be on an organized high-school rodeo team, well, that was amazing. We had these red rodeo shirts with NMMI patches on them, and we traveled together in an NMMI bus while Boss and his wife Mabel brought the horses in his rig.

Mabel wasn't her real name, but I had hung that handle on her and it seemed to stick. She sure did look like a Mabel. Our other rodeo coach was a former jockey named Mike. He knew a lot about horses, but I don't remember him knowing all that much about rodeo. I could be wrong about that though.

My kids have since competed in that arena, but it had changed from the days that I remember. They no longer allow people on the rocks around the arena, so the feeling is different from the one that we used to get riding into the arena. Those Indians on the cliffs in my day were simply rodeo fans, but you could feel like you were riding into an ambush. It was such a cool place to throw down a steer.

My folks went to every rodeo that they could make it to. Mom would pack food for everyone. Our area of the bleachers was certainly a popular spot among the rodeo team members who had been existing on the offerings from the Den of Culinary Atrocities.

We had the final Annual Formal Inspection in April, went streaking on a much smaller scale because the National Guard was called out to prevent our fun, and we finished up the rodeo season in style. Most of the team competed at the New Mexico High School Finals Rodeo, with a bunch of us qualifying for the National High School Finals in Rapid City, South Dakota.

I had continued in my position as the mounted representative on the dismounted Color Guard, so I probably went to fewer formations in the last half of my senior year than any other cadet that year. I was skating along and shirking my military responsibilities. Not something that I am proud of, but it is what it is. My goal was to rodeo, and I was going to go to college only so I could rodeo in college. From my seventeen-year-old perspective, that looked like a great career. No wonder people are not legally adults until after they are eighteen.

Graduating after three years at NMMI put me in big demand at colleges in the southwest. Those colleges were aware of the things that a cadet had gone through to achieve a diploma. Having been on a great rodeo team and qualifying for both state and national finals as a steer wrestler also helped. There were several full-ride scholarships offered, but I ended up taking one from Allen County Community College in Iola, Kansas.

I thought I was moving there to live in an arena owned by one of my rodeo heroes. What I didn't know was that God was sending me there to meet my soul mate and dash my rodeo dreams as I transitioned from rodeo cowboy to farrier.

NMMI continues to enjoy a fantastic reputation among schools, whether military schools or not. The stringent academic standard and constant physical demands create stronger adults. It is incredible how many of the graduates from NMMI have gone on to great success in many different areas. I believe much of that success can be attributed to having gone through and survived being a cadet, and I am thankful to be an alumnus of the New Mexico Military Institute.

Graduation in 1986 was fantastic. There was a military parade as part of the ceremonies, and I proudly rode my horse at the head of the Mounted Color Guard. We were told not to throw our caps into the air, but we did it anyway since we no longer fell under the command of those in power. It seemed like such a rebel act and signified for me the end of that time of my life.

I have gone on as an educator to use many of the examples I learned as a cadet. People say that learning at Heartland Horseshoeing School is like drinking from a firehose. NMMI felt like that in many ways. I keep the pressure turned up to maximum on my students, and it produces results.

My career has been one of setting unrealistic expectations for people, then fully expecting they meet them. Some people hear about how hard the Heartland is and say, "Man, that place is not for me." Those folks are right about that. Then there are ones who say, "Bring it on. I'll test myself against and amongst the best." Well, they are also right. NMMI was exactly that way, as well. Those who thought they were not tough enough or determined enough to get through, well, they were correct.

New Mexico Military Institute will always be a special place for me. It was a forge that formed, hardened, and tempered my person so that I could go on from there and do those things that I wanted to do.

49

FINAL DAYS AT THE RANCH

Little Mike and I headed to Rapid City, South Dakota for the 1986 National High School Finals Rodeo. It was massive and a lot of fun. The experience only solidified in me the desire to make rodeo a career, and I did not understand how anyone would not want to be a rodeo cowboy.

The finals did not go as hoped. There are a thousand other cowboys who had a similar experience, but only the winners thought it was the best finals ever. That story repeats itself, at a different scale and format, in every sport in the world. However, disappointment leads to working harder and longer and builds those who have the will to be greater than they are.

"I'm headed to Kansas next week," I told Little Mike. "I can't wait for college rodeos."

"You're lucky," he said.

I nodded in agreement. "That rain was something else," I said.

We were sitting on the porch of the ranch house. The finals were behind us, and I was enjoying the last few days at the ranch I grew up on.

"Wasn't nearly as much as that time you sent us down the river on that raft," Little Mike observed.

"No," I said. "You still remember that?"

"Are you kidding?" he asked. "My shoulder still hurts from that little adventure."

"Sorry, man."

"Are you still going on your date?" he asked.

"Of course," I said.

"Dead Horse is bound to be up," Little Mike advised, referring to the ditch we had to cross known as Dead Horse Creek.

"I can't imagine it'll be a problem," I said.

Little Mike only smiled.

For 360 days of the year, Dead Horse did not have enough water in it for a grasshopper to get a drink. When a storm rolled through the mountains though, well, it could become quite the crossing.

I prettied up the best I could with what the Lord had graced me with and pointed my Chevy toward town. About two miles in, I came to the banks of Dead Horse Creek, which were about one hundred feet farther west than they had been before the storm. I was at that eighteen-year-old place in life where no one had ever been before me. No mountain too high to climb, no horse too rank ride, no steer too strong to throw, and certainly no river too wide to cross. Especially when I was in my new truck that shared a similar sentiment to mine. My feet could feel its power and desire through the floorboards.

My main decision was whether to hit the water crawling or at speed. Growing up in New Mexico meant that my overall experience with water crossing was minimal. Even though I can look back from here and see that the experience I did have should have been sufficient, I wasn't seeing that at the time.

I decided speed was the answer, as you have probably already guessed. I backed up, shifted into first, and mashed the gas as I popped the clutch. The tires spun fast, swerving and slinging mud before catching in the wet sandy clay leading down to the torrent. Skidding as I gained control, I hit the muddy water. Momentum carried me further than traction, but soon the truck was moving sideways instead of forward.

The direction of the stream meant that the water was pressing on the driver's side. My truck and I were floating to the right, the water at least halfway up the door. I rolled the window down and jump/swam to the far side of the creek. A lucky tree had prevented my truck from heading toward the Rio Grande.

Our nearest neighbors were RD's place. He lived about a half mile from where I had tried to see if my truck was also a boat. I walked to RD's house, and he graciously helped me get my soggy truck out of the water.

I made it back to the ranch and we got the truck to the dealership the next day where they worked on it and repaired whatever damage I had caused. It was a general pain, and I ended up not going on my date.

This was the end of an era for me, even though I didn't know it at the time. My plans were to go to college, go to vet school, and someday return to the mountains of New Mexico and open a practice, all the while rodeoing every weekend and competing at the Professional Rodeo Cowboys Association National Finals Rodeo every year. At that point in my life, the rodeo was more important than school, and I obviously had no idea what the demands of vet school would have required.

I left New Mexico in the last part of July to begin the adulthood part of my life. It was off to Kansas where I had an arena to live in and classes to go to that were being paid for by a rodeo scholarship. I did live in my parent's house for about six weeks the following summer, but the reality was that from July of 1986, I was no longer a citizen of New Mexico.

"Well, I'm off," I said to my mom and dad.

It was just after sunrise, and we were sitting on the front porch of the ranch house. Mom had just made a great breakfast and handed me a bag of food for the road. My green and white two-horse, gooseneck trailer with a tack room was hooked up to the little beige and white Chevy that had been "fixed" from trying river rafting.

"How long a trip are you expecting?" Dad asked.

"I think it will be a little over 12 hours," I said. "Depends on if I can get behind a trucker."

The maximum speed limit in the US at that time was only 55 miles per hour. It seemed so slow that you could get out and run along beside the truck, but if you got behind a semi, they had speed detectors and CB radios, so they were often at least ten miles over the limit.

"Please be careful, *mijito*," Mom said, giving me a hug. "Call us when you get there."

"Will do."

I kissed my mom, shook my dad's hand, and headed to my truck. I crossed the cattleguard without knowing what lay ahead or what God had planned for me, but I cranked up the Chris Ledoux cassette tape on the stereo and drove down the dirt road toward my destiny.

It has been one fantastic journey.

ABOUT THE AUTHOR

Chris and his family have made a career of shoeing horses and teaching people this craft. He lives on a farm in Missouri with his wife Kelly. Both their kids and their families live close are involved in shoeing, running Heartland Horseshoeing School, and their respective farms. Chris and Kelly love watching their grandkids grow.

Made in the USA
Monee, IL
04 November 2026

67381185R00125